Fine Homebuilding®
on
Frame Carpentry

Fine Homebuilding®
on
Frame Carpentry

The Taunton Press

Cover photo by Paul Spring

©1990 by The Taunton Press, Inc.
All rights reserved

First printing: March 1990
International Standard Book Number: 0-942391-53-5
Library of Congress Catalog Card Number: 89-40576
Printed in the United States of America

A FINE HOMEBUILDING Book

FINE HOMEBUILDING is a trademark of The Taunton Press, Inc.,
registered in the U.S. Patent and Trademark Office.

The Taunton Press, Inc.
63 South Main Street
Box 5506
Newtown, Connecticut 06470-5506
U.S.A.

C O N T E N T S

INTRODUCTION

Building a proper frame makes more than a sturdy and durable home. It makes for plumb walls, level floors and trimwork that fits precisely. This means that even the simplest frame must be accurately laid out and built with care. And it means that complex jobs, like framing hip roofs and dormers, must be done with confidence that comes from patient study or from long experience. Sometimes from both.

This book brings together 27 articles about frame carpentry from the back issues of *Fine Homebuilding* magazine.* We hope you'll be able to use it the next time you're about to crawl up on a roof to noodle through a framing problem, and you want informed advice from another builder who's been there before.

John Lively, editor

*The six volumes in the *Fine Homebuilding* on... series are taken from *Fine Homebuilding* magazine numbers 1 through 46, 1981 through early 1988. A footnote with each article tells when it was originally published. Product availability, suppliers' addresses, and prices may have changed since then.

The other five titles in the *Fine Homebuilding* on... series are *Builder's Tools; Foundations and Masonry; Doors and Windows; Floors, Walls and Stairs;* and *Baths and Kitchens.* These books are abridgements of the hardcover *Fine Homebuilding* Builder's Library.

Building Basics

What makes structures stand up, and fall down

by Thomas Koontz

I have always been fascinated by Frank Lloyd Wright's Fallingwater (photo below). Delicately balanced on concrete piers and anchored into a stream-side boulder, its concrete balconies cantilever over the waterfall below. In apparent defiance of the laws of nature, terraces, layered one on top of another, appear to be supported only by the glass and horizontal mullions of the windows below them. Given nature's dynamic forces, it is difficult to understand how such a form, so visually and spatially powerful, could be structurally stable. But it is.

Like Fallingwater, all construction must take into account the laws of nature. It doesn't matter what structural gymnastics we attempt to make it appear otherwise. Gravity, wind, ground movement and changes in temperature and humidity all generate internal and external loads

Thomas Koontz is assistant professor of architecture at Virginia Polytechnic Institute in Blacksburg. He is also a practicing architect.

on our buildings. The structural systems of buildings must be designed to perform under these loads. The success of Fallingwater's structure, and of any building for that matter, is due to the strength of its materials and to their proper behavior in assembly under stress.

To know whether our materials and systems will be strong enough and will behave as they're supposed to, we must first know what loads will be exerted upon them. When I speak of loads, I refer to the distribution of a building's *live loads* (people and their furniture) and *dead loads* (the weight of the structure itself), as well as of the environmental forces imposed upon it. Our first task in determining the loads is to establish their direction (horizontal or vertical) and their magnitude (degree of force). Once we know how they will be loaded, structural components can then be selected and the assemblies designed.

Compression and tension—Just as the laws of nature have not changed, neither have the

ways structural components behave. We may construct the components in different ways and of different materials than in the past, but they must still withstand the same old forces. It was Isaac Newton who first formulated our understanding of how natural forces behave. Newton's Third Law of Motion states that for every action or force there is an equal and opposite reaction or force. Accordingly, when we consider a load pushing down on a column, we realize that the ground upon which it bears must push up with the same force (figure 1, below). If the ground didn't push back with equal strength, the column would slowly sink. (Unfortunately, we all know examples of this.) The load and the ground both push against the column, compressing it between them, hence the term *compression*. Reverse this, pull the column, and the ground will also react by pulling (figure 2). This action will attempt to lengthen the column and place it in *tension*. Structural assemblies do all their work by these two actions—compression

Fig. 1: Compression
Compressive forces act to shorten the material.

Fig. 2: Tension
Tensile forces act to stretch the material.

Frank Lloyd Wright's Fallingwater, pictured here with some of its shoring still in place, is an excellent example of how structural laws can be stretched to their very limits. The final illusion of heavy concrete slabs hovering in the air above the stream shows what can be done, given the right materials and canny engineering. Photo courtesy of the Avery Library, Columbia University.

From *Fine Homebuilding* magazine (April 1986) 32:29-33

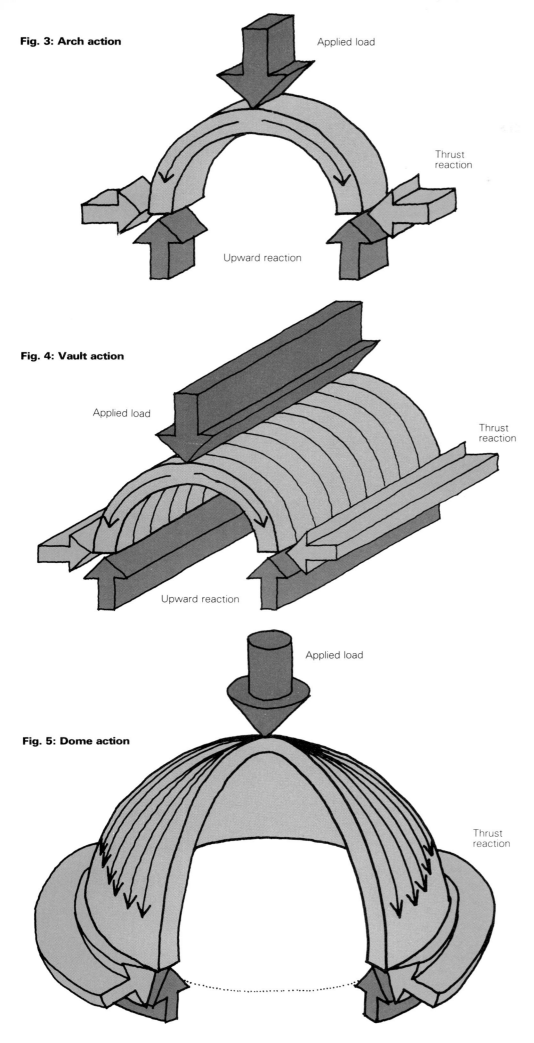

Fig. 3: Arch action

Applied load

Thrust reaction

Upward reaction

Fig. 4: Vault action

Applied load

Thrust reaction

Upward reaction

Fig. 5: Dome action

Applied load

Thrust reaction

and tension. Buildings and their materials are either pushed or pulled by the loads exerted upon them.

Arches, vaults and domes—We have been building in compression ever since we placed one stone on top of another. Since then, simple walls and columns of stone or wood have become increasingly complex. With the use of the arch, large openings in walls could be spanned in compression. The arch (figure 3, at left) relies entirely on compression; each stone or brick pushes outward and is resisted by the mass of the wall. By lining up one arch beside another, the arch became a vault (figure 4), and the use of stones was no longer limited to building walls. The arch could now be used for roofs and ceilings. Rotating a series of arches in plan from a center point created a dome (figure 5), which fully enclosed the overhead plane. The vault and dome, just like their predecessor the arch, require a mass to counteract the thrust of their loads. Chains were used to encircle the base of the dome and thereby restrain its tendency to spread outward under its own weight.

The vault was restrained by thickening its supporting wall or by adding arched buttresses to transfer the outward, horizontal thrust down to the ground. The architecture of compression reached its peak in the Gothic cathedrals. The entire cathedral of stone was in compression, held together by gravity.

What ultimately held up the cathedral, the foundation, also holds up our houses. The foundation distributes the loads of the building to the ground. And, by Newton's Third Law, the ground had better react by pushing upward against the foundation with an equal force. The foundation/ground reaction may take place through bearing, as with the spread footing and or rubble trench; by friction, as with the pile foundation; by spanning, as with the grade-beam foundation; or by floating, as with the raft foundation (slab on grade). These types of foundations all behave by pushing or pulling.

Bending, shear and deflection—The simple forces of pushing and pulling often combine to generate forces other than simple tension or compression within materials and assemblies. For example, a masonry block fails when it cracks and finally fractures. Interestingly, the fracture takes place when energy builds up within the block at the molecular level along certain distinct lines. As one molecular layer is compressed and its adjacent layer is simultaneously tensed, a slipping action takes place between the layers. This ultimately causes the layers to separate and fracture. This phenomenon, called *shear*, happens when two forces act in opposite directions and in adjacent planes (see figure 6, next page).

Another phenomenon created by co-active tension and compression is *bending*. Beams provide the most common examples of this. By intuition we know that a loaded beam will bend in the middle. The material in the top half of the beam is shortened as the beam's concave curve pushes it inward and compresses it (figure 7). The bottom of the beam is put in tension as the

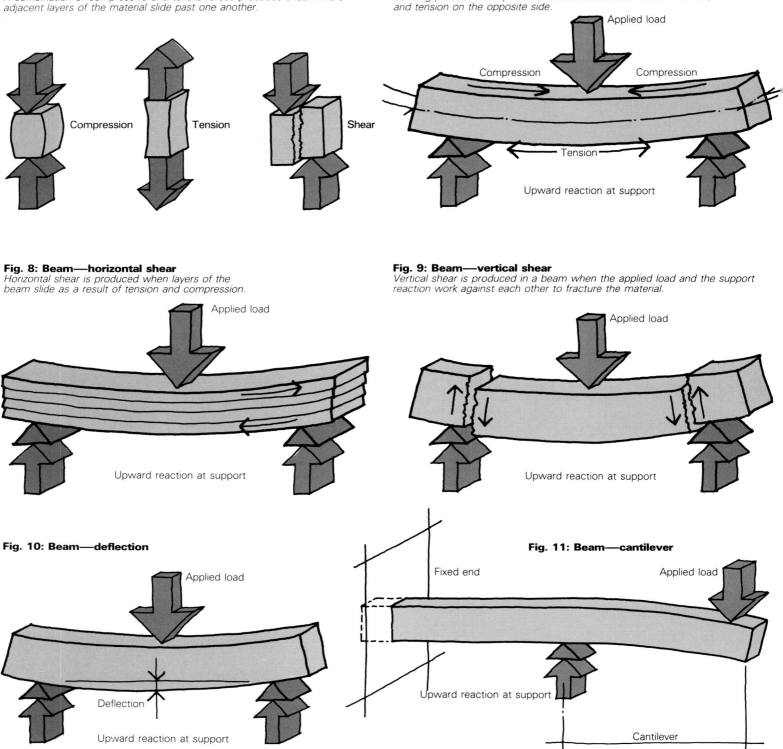

Fig. 6: Shear
A combination of compressive and tensile forces produces shear where adjacent layers of the material slide past one another.

Compression   Tension   Shear

Fig. 7: Bending
Bending produces compression on the loaded side of the member, and tension on the opposite side.

Applied load

Compression   Compression

Tension

Upward reaction at support

Fig. 8: Beam—horizontal shear
Horizontal shear is produced when layers of the beam slide as a result of tension and compression.

Applied load

Upward reaction at support

Fig. 9: Beam—vertical shear
Vertical shear is produced in a beam when the applied load and the support reaction work against each other to fracture the material.

Applied load

Upward reaction at support

Fig. 10: Beam—deflection

Applied load

Deflection

Upward reaction at support

Fig. 11: Beam—cantilever

Fixed end   Applied load

Upward reaction at support

Cantilever

convex side of the curve lengthens and stretches it. The central axis of the beam is neutral, as the material is being neither pushed nor pulled.

Shear exists in bending, both horizontally and vertically. Internal forces attempt to shear the beam horizontally (parallel to the neutral axis) as the molecular layers of the material are pushed and pulled in different planes (figure 8, above). Vertical shear develops as the beam and its load push down and the support upon which it rests reacts by pushing up (figure 9). Again we have two forces opposing each other in different planes. We can expect vertical shear to develop in the beam just to the inside of the beam

supports. As we shall see, wood and steel both behave well in tension and compression and are able to resist shear forces proportional to their strengths. Concrete, however, while holding up well in compression, requires steel reinforcing to resist shear successfully.

Generally, the material, size and configuration of a beam are not determined by the beam's failure in bending or shear. Instead, the size is often established by its *deflection*. Deflection is the measure of how much the beam bends under load; it is the dimensional change that takes place when the beam is stressed (figure 10). Deflection is typically specified by the building

codes, where it is not allowed to exceed $\frac{1}{360}$ of the span. Most beam designs reach the allowable deflection limit long before they reach their limit in bending. So although we need to determine the bending and shear stresses in the beam, it is most often the deflection that governs the size of the beam.

The cantilevers of Fallingwater are beams that are free at one end while supported at the other end and in the center (figure 11). Because there are no columns or walls, the terraces have an uninterrupted flow of space and view. The bending and shear forces in a cantilever are not equally distributed between two supports. They are concentrated at one support and so are greater than a simply supported beam.

Anticipating the internal stress patterns allows us to design a beam that is structurally and materially efficient. The combination of the tension, compression and shear forces in the beam makes a simplified pattern of Xs expressing the resulting forces flowing through the beam. These Xs can be delineated and fabricated of materials that are strong in either tension or compression, whichever applies, hence the development of open-web joists and trusses (figure 12, below). Trusses can span greater distances with less material and weight than solid beams because their geometries match the anticipated stress. The top chord acts in compression; the bottom chord acts in tension; and the web members take up the shear and the co-active tension and compression. (For more on trusses, see pp. 27-31.)

Walls and columns—Although walls and columns are essentially compressive elements, if one is loaded eccentrically (the load is not centered) or if the load becomes too great, the wall or column may buckle. Buckling causes the member to behave like a beam by bending it out, placing it in tension as well as compression. Resistance to buckling is a function of the member's thickness in its weakest dimension, relative to its unsupported height. The possibility of buckling is reduced by thickening the member or reducing its effective height by bracing it about two or more of its axes.

Walls are inherently stable parallel to their length. They are subject to bending from lateral loading, however, across their width. This is recognized by the building codes when they specify thickness-to-height ratios for different wall materials. Horizontal loading of walls from earth retaining, wind or earthquakes raises the possibility of overturning. Placing pilasters or cross walls at specific distances relative to the height, thickness and kind of material of the wall stiffens the wall and resists overturning. Again, the

codes recognize this and set minimum standards. The north retaining wall of the Jacobs II house in Wisconsin by Frank Lloyd Wright uses the form of an arch, in plan, to resist the horizontal load of the earth.

A wall that retains earth acts either as a vertical cantilever or as a beam. The cantilever type requires a fixed foundation condition that will resist overturning. This is usually achieved by building an oversized footing on the loaded side. The load then helps to restrain the footing and thus reduce the possibility of overturning. The wall is then designed as a vertical cantilever, fixed at the base and free at the top. Just as in a horizontal cantilever, bending and shear stresses must be considered (figure 13, p. 12).

The beam-type retaining wall resists the lateral load of the earth by spanning between rigid vertical members such as pilasters or cross walls. The wall distributes the load to these members just as a beam would distribute its load to its supports (figure 14). Similarly, the wall must be designed for bending as well as for compression. A basement wall is a good example of a retaining wall that behaves as a beam between the end walls or cross walls. The loads of the floors, walls and roof above place the wall in compression. This compression load actually strengthens the wall by making it more rigid in response to the bending forces.

The compressive nature of foundations can also be affected by bending. Different rates of compaction of soils can create voids in the soil, and the foundation can settle unevenly. As the footings attempt to span the voids, they begin to bend and behave like beams. This creates tension in the otherwise compressive footing. Since we usually construct our footings out of concrete, and concrete is weak in tension, reinforcing steel is placed in the concrete to resist the buildup of tensile forces.

Knowing the properties of materials helps us to make decisions on what to use. Understanding the behavior of loads and materials lets us know why the concrete footer will crack when

bending; and why we were smart to use the steel to reinforce it. Some materials are good in tension, some in compression and some in both. The strength and behavior of materials under stress are critical to the successful performance of the structure.

Behavior of materials—Materials that can strain, change shape or deform and then return to their original shape are said to behave *elastically*. Elasticity is an important property since it allows the material to assume its original shape after the load is removed.

Materials have elastic limits, however, and can be stressed beyond them. This causes permanent deformation in the material, and the original shape is lost. Once the material deforms permanently, it is said to behave *plastically*. For example, a steel beam will bend if too much weight is placed on it. Even after the load is removed, the material remains deformed. Failure usually occurs once the elastic limit of the material has been exceeded. Materials that are *brittle* do not perform elastically. They have little or no ability to deform and reform and thus can fail without warning.

Structural materials need to be *ductile* (not brittle), elastic, strong, and have the ability to be pushed and pulled. They must exhibit all or most of these attributes to remain stable in the real-world environment. Manufacture, growth characteristics, fabrication and geometry contribute to the structural properties of materials.

Wood is an organic material, and it grows by cellular division into long fibers. These fibers transport food up and down the entire length of the tree and give wood its characteristic strength. Consider an analogy: a bundle of drinking straws bound together. The straws may support 20 lb. or more if they are loaded vertically, parallel to their length. If they were equally loaded perpendicular to their length, however, it is easy to see that the straws would fail to support the load. So it is with the fibrous nature of wood. It is stronger parallel to the grain than perpendicu-

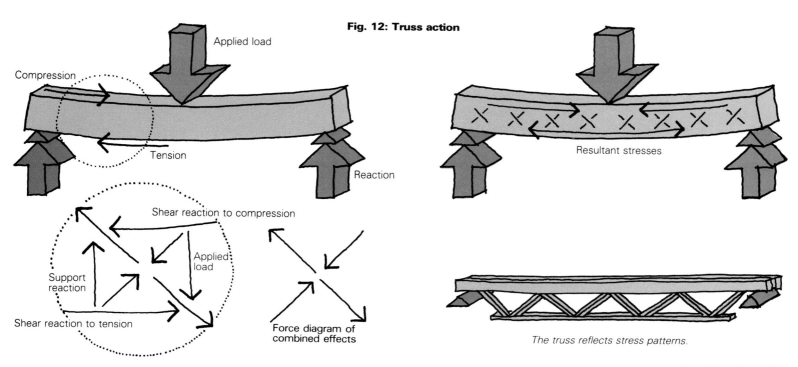

Fig. 12: Truss action

Applied load

Compression

Tension

Reaction

Shear reaction to compression

Support reaction

Applied load

Shear reaction to tension

Force diagram of combined effects

Resultant stresses

The truss reflects stress patterns.

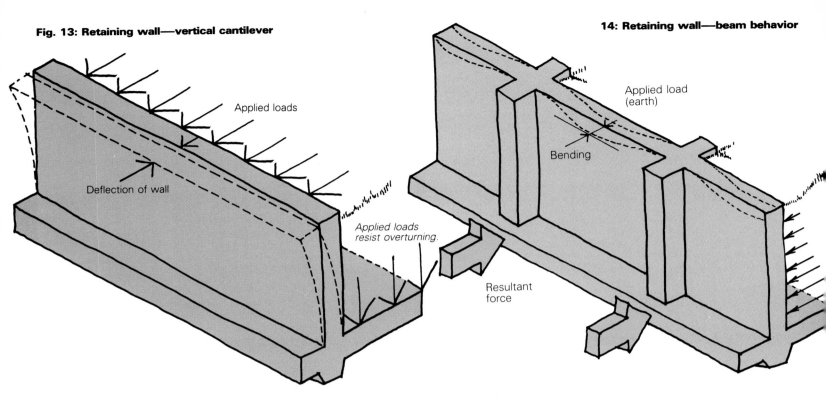

Applied loads

Deflection of wall

Applied loads resist overturning.

Resultant force

Applied load (earth)

Bending

lar to the grain. Its fibrous character performs well in tension and compression, although it is stronger in compression.

Relative to other traditional structural materials, wood is softer and weaker. Its strength is a function of its species, seasoning and growth characteristics. The compressive strength of wood parallel to the grain ranges from as low as 400 psi for balsam fir up to 1,850 psi for Douglas fir. Many species of wood exhibit dimensional instability under mechanical, thermal or moisture stress. Wood tends to expand or contract, thereby placing itself and its joints under additional stress and strain. The popularity of wood is due to its availability, workability, manageable size and weight, and its structural capacity for light construction projects. It can be milled into a variety of cross-sectional shapes, making it useful for many structural applications. Because it performs well in both tension and compression, wood is a natural for bending members.

Plywood, because it is made up of thin laminae glued one atop another, is dimensionally stable and resists tension and compression well parallel to its plies. So it is well suited for structural applications such as truss webs and shear panels. But plywood bends easily when forces are applied perpendicular to the plies, and so has poor spanning characteristics.

Steel, on the other hand, is a hard, dense crystalline solid, with high elastic and tensile strengths. Strong in tension and compression, steel can be used in almost any structural application where corrosion, weight and fabrication difficulties are not problems. Columns, beams, joists, trusses and even composite assemblies are fabricated with steel. To compensate for their lower tensile strengths, concrete, masonry and wood are often combined with steel.

The carbon content of steel determines its hardness and ductility. The greater the carbon content, the harder the steel, but the less ductile it is. Plain carbon structural steels are composed

of 0.15% to 0.25% carbon. If the steel contains less carbon than this, it becomes soft and malleable and is classified as wrought iron. If it contains more than 1.7% carbon content, the steel becomes brittle and is called cast iron. The approximate relative tensile strengths of these types of steel are as follows:

> Wrought iron: 48,000 psi
> Mild steel (A36): 36,000 psi
> Cast iron: 22,000 psi

In comparison, aluminum's tensile strength is around 11,000 psi.

Concrete is a composite made up of cement, water and aggregate. It is strong in compression, but its tensile strength is only about 10% of its compressive strength. It is hard and has a low elastic limit. Over time it has the tendency to *creep,* and may shorten or stretch under constant stress. The strength of concrete is based on aggregate size, ratio of water to cement, and ratio of cement to sand. Compressive strengths range from 2,500 psi to 7,600 psi. The highly plastic nature of its installation makes it unique in the field for its formability.

Many of concrete's structural weaknesses can be overcome through the addition of steel. Reinforcing concrete with steel reduces its shrinkage and creep, and steel increases its bending capacity. All this is achieved because of steel's high tensile strength. Concrete is best suited for compressive loading, but with the addition of steel, it can be successfully used in members subject to bending and shear.

The terraces of Fallingwater are like large reinforced concrete trays with turned-up edges. Cantilevered from stone walls, boulders and concrete piers, the terraces are stressed in bending and shear. The walls and piers act as columns and the terraces as beams. The reinforced concrete provided Wright with the plasticity in construction and the compressive and tensile strengths required to resist bending.

Like concrete, unit masonry is a compressive

material. It also has relatively little tensile strength and a low elastic limit. Any tension that exists in masonry members must be handled by the addition of steel reinforcing. Purely compressive forms like columns, walls, arches and vaults are appropriate for masonry construction. Masonry strengths are a function of the unit material and the manufacturer. Clay masonry has compressive strengths of 1,500 psi to 3,000 psi, on the average. Unless they are vitrified (kiln burned to reduce their moisture-absorptive nature) bricks and other clay masonry units are subject to dimensional instability when they get wet. Concrete masonry shares this propensity to moisture stress. The strength of concrete masonry depends upon its aggregate type and size, mix and consistency and falls between 700 psi and 1,800 psi, on the average. Stonemasonry's compressive strength is based on the stone's formation process. The following are representative examples of the relative compressive strengths of stone:

> Granite: 15,000 psi to 30,000 psi
> Slate: 10,000 psi to 23,000 psi
> Marble: 9,000 psi to 17,000 psi
> Limestone: 3,000 psi to 21,000 psi

The type and strength of mortars, workmanship and materials all contribute to the structural stability of masonry assemblies.

Frank Lloyd Wright used masses of stone masonry to counterbalance the cantilevered terraces at Fallingwater. The strongly horizontal courses of stone, layered three stories high, not only stabilize the cantilevers but also create a sense of refuge and quiet from the more dynamic terrace spaces. The stone provides a texture and warmth that the concrete cannot. It also provides an expressive counterpoint of compression to the bending of cantilevers.

Although Fallingwater is a complex and visually exciting structural exhibition, it remains standing because its architect understood the inherent nature of its materials.   □

Wood Foundations
Pressure-treated studs and plywood make an economical system for owner-builders

by Irwin L. and Diane L. Post

We first heard about All Weather Wood Foundations (AWWF) in 1978. We later learned more about this pressure preservative-treated system, and when we began to build our own house we chose it over the more common poured concrete and concrete-block foundations. There were several reasons for our choice.

First, the rugged winters here in the mountains of Vermont made the insulation of the house one of our primary concerns. Since we planned our basement to be living space, we wanted it to be as well insulated as the rest of the house (fiberglass to R-26 in the walls and R-38 in the top floor ceiling). AWWF walls are built of studs, and they can be insulated with fiberglass as easily as can any stud wall. With concrete foundations, a thick layer of insulation significantly reduces the usable space inside the basement, and it's difficult to attach insulation and the interior finished wall to the concrete. The stud walls of the AWWF allowed us to hide the wiring and plumbing, too.

The AWWF also suited our construction schedule. We were able to start on it as soon as the excavator finished the cellar hole. For a poured foundation, we would have had to hire a contractor and wait for him to work our job into his schedule. We would then have had to wait for the forms to be erected, and the concrete to be poured and to cure. For a concrete-block foundation, we would have had to pour a concrete footing and then begin the time-consuming task of laying the blocks.

The AWWF facilitated the installation of windows and doors. We simply nailed and screwed them in, as in ordinary frame construction. An error in pouring a concrete foundation can be disastrous when it comes to installing doors and windows; just a little reframing corrects an error with the AWWF. We decided to relocate a door slightly—it took us only two hours.

We wanted to use spruce clapboards for our exterior siding, and we would be able to nail the clapboards directly to the exposed portion of the wood foundation. We're not sure what we would have done with the above-grade portions of a concrete foundation.

The clincher for us was the money we saved. We were the labor force for everything but the excavation, standing-seam steel roof and drywall. Like most people who build their own homes, we did not include our labor as part of the cost. While we didn't do a detailed analysis for all the options, we estimated a savings of more than $1,000 over the cost of poured concrete. Materials for our 24-ft. by 32-ft. foundation cost us about $1,850 in August 1980.

Many people ask us if we're worried about the foundation rotting out. We aren't. The required preservative salt retention in the pressure-treated wood is 0.60 pounds per cubic foot, which is 50% higher than building codes require for general ground-contact applications. The USDA Forest Service's *Wood Handbook: Wood as an Engineering Material* indicates that test stakes in Mississippi have lasted more than 20 years at lower preservative salt retentions.

The AWWF is a stud wall sheathed with plywood on the outside. The walls stand on footing plates, which lie on a pad of gravel. A concrete slab poured inside the walls prevents the backfill from pushing in the bottoms of the walls. The tops of the foundation walls are securely fastened to the first-floor structure before the backfilling begins.

The size of the footings and studs and the thickness of the plywood depend on the size of the building, the grade of lumber, the depth of the backfill and the spacing of the studs. An industry booklet, "The All-Weather Wood Foundation: Why, What and How" ($1 from the American Plywood Association, Box 11700, Tacoma, Wash. 98411), supplied enough information for us to design our AWWF with confidence. We used 2x10 footing plates, except at the back wall, where we used 2x8s because of the smaller load on the back wall. Our design called for ½-in. plywood and 2x8 bottom plates, studs and top plates. The top plates and the plywood that was more than 1 ft. above grade were not preservative treated. We choose to use 2x8 studs on 16-in. centers so we could fit two layers of R-13 insulation inside the exterior walls. Our house design required very low-grade studs (F_b of 975 psi minimum) for adequate strength with our depth of backfill (up to 5 ft.). The fasteners were 10d stainless steel nails to connect the bottom plates to the studs and footing plates, 8d stainless steel nails to connect the treated plywood sheathing to the studs below grade, 8d hot-dipped galvanized nails to connect the untreated plywood to the studs above grade, and

Once the gravel pad is compacted and carefully leveled, left, the footing plates are set in place around its perimeter. The layout of the excavation and the drainage pipes is shown in the drawing on the facing page. Right, pressure-treated foundation framing is built 8 ft. at a time and tilted up. The interior bearing wall is framed with less costly untreated lumber because it won't be in contact with water or wet earth. Photos: Irwin and Diane Post.

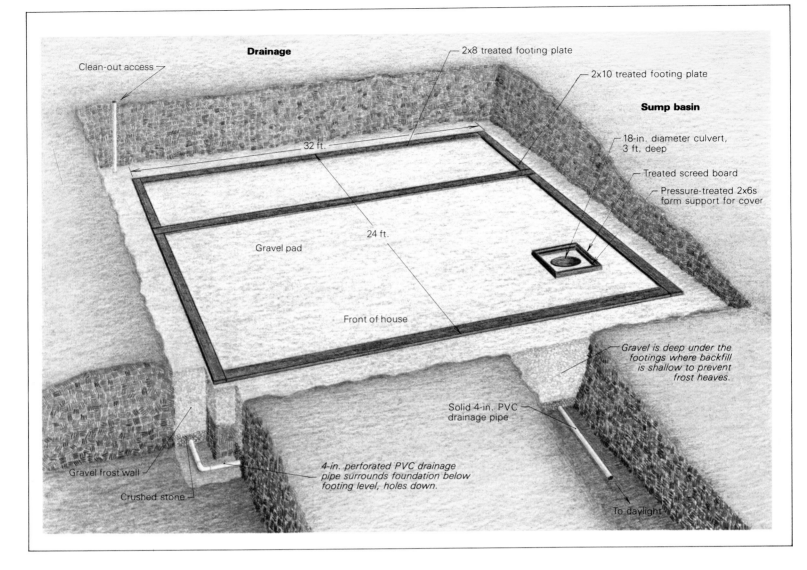

Drainage

Clean-out access

2x8 treated footing plate

2x10 treated footing plate

Sump basin

18-in. diameter culvert, 3 ft. deep

Treated screed board

Pressure-treated 2x6s form support for cover

32 ft.

24 ft.

Gravel pad

Front of house

Gravel is deep under the footings where backfill is shallow to prevent frost heaves.

Solid 4-in. PVC drainage pipe

4-in. perforated PVC drainage pipe surrounds foundation below footing level, holes down.

Gravel frost wall

Crushed stone

To daylight

16d hot-dipped galvanized nails to connect the top plate to the studs.

We found two lumberyards that were willing to bid on our AWWF materials. The better bid was for far higher-quality studs than we needed and for the ⅝-in. plywood the supplier had in stock rather than the ½-in. plywood we'd specified. We could have saved if we had been willing to wait for material closer to our specifications. The price for the stainless steel nails seemed unbelievable at $6 per pound. This works out to about 10½¢ for each 10d nail!

In ordering our AWWF materials, we specified that the wood had to be stamped with the American Wood Preservers Bureau (AWPB) foundation grademark, which ensures that the wood is properly treated for use in foundations. We were pleased to find many plugged holes in our material where samples of wood had been removed after treatment to check for retention of the preservative.

Excavation and drainage—We had two excavators bid on our job. Neither one had ever worked on a cellar hole for an AWWF. After reviewing our engineering drawings and instructions, one excavator seemed reluctant. The other showed interest, so we chose him.

The main objectives of the excavation were to lay the gravel pad on which the footings sit and to provide good drainage around the foundation. Good drainage is an absolute necessity for any foundation. After having suffered with wet basements, we were not about to take any shortcuts with our new house. The design we settled on is diagrammed above.

In the front and sides of the cellar hole, where the footings were going to be less than 4 ft. below finished grade, we wanted a frost wall built to prevent frost heaving under the footings. We had a ditch 2 ft. wide dug to about 5 ft. below finished grade, and set drainage pipes surrounded by crushed stone in the bottom. The ditch was then backfilled with gravel.

Along with the frost wall, we had a sump basin excavated inside the foundation walls. This basin is simply a hole in the basement floor connected by pipe to the drains around the house. Groundwater normally flows to the drain-pipe outlet. If the outlet becomes plugged, the water backs up into the sump basin so we can pump it outside, keeping our basement dry. We used a 3-ft. section of corrugated aluminum culvert 1½ ft. in diameter to form the sump basin. A treated wood cover over its top is flush with the concrete slab.

Next, we worked on the gravel pad. It varied in depth from 6 in. (the required minimum) to more than a foot. We used crusher-run gravel, which does not contain large cobbles. Pea-stone,

coarse sand or crushed rock could have been used instead. The excavator drove his bulldozer back and forth over the gravel to compact it. We made sure the pad was large enough to lay out our footings at 24 ft. by 32 ft., and we marked the locations for the footing plates.

After the gravel was compacted, we leveled the pad by driving 2x2 stakes of treated wood into the gravel at 6-ft. intervals along the critical lines where the footing plates were to lie. Using a surveyor's level for accuracy, we drove the top of each stake to exactly the same elevation, which was close to the average elevation of the gravel. Then, using shovels and garden rakes, we leveled the entire pad to match the tops of the stakes. To level the footing lanes exactly, we scraped a straight 2x4 stud over the gravel between the stakes. Leveling took half a day and some patience and care.

Next we installed the ABS wastewater pipes that were to run under the basement floor. We laid out their positions precisely on the gravel surface, then we dug the ditches, put the pipes in place (gluing the joints well), covered them, and releveled the disturbed areas. We were then ready to set the footings.

Framing—The footing plates were simply laid flat on the ground (photo, previous page, left). We made careful diagonal measurements with a

Back and interior wall design

- Double 2x4 stud wall
- ½-in. sheathing
- Clapboards over untreated building paper
- 2x3 thrust plate
- Three 2x8 top plates
- Caulk
- Treated board to protect top of polyethylene sheet
- Native soil backfill
- 6-mil polyethylene moisture and vapor barrier
- Two layers of R-13 fiberglass insulation
- 2x8 treated studs
- ⅝-in. treated plywood
- Gravel backfill
- 2x8 treated bottom plate
- 2x8 treated footing plate
- 4-in. perforated PVC drainage pipe
- Crushed stone
- Gravel pad (6-in. minimum)
- 2x10 treated footing plate
- Two 2x8 treated bottom plates

- ½-in. plywood subfloor
- Two 2x6 top plates
- 2x6 untreated studs
- 2x10 joist
- Drywall
- Drywall
- 4-in. concrete slab
- Treated screed board

fiberglass tape to ensure that the footings were positioned squarely. Sections were cut out of the plates to accommodate the wastewater pipes located in the bearing walls.

There were just the two of us, so we were not able to handle long, heavy wall sections. We framed one 8-ft. section at a time, stood it in place, nailed it to the footing plates, and nailed on the top plates so as to connect adjacent wall sections. As the wall took shape, we frequently checked for plumb with a 4-ft. carpenter's level and a plumb bob.

The drawing above contrasts the design differences between the back wall and the interior bearing wall in our foundation. We did not use preservative-treated wood for the interior foundation bearing wall of 2x6 studs. By adding an extra bottom plate of treated wood and trimming off the appropriate length from each stud,

we raised the bottom of the studs above the level of the concrete slab. Using untreated studs in this wall saved us a lot of money.

When the foundation walls were all in position, we nailed sheathing onto their lower halves and fully sheathed some of the corners to stiffen the structure. Then we brushed the cut ends of the foundation wood with a generous coat of preservative, and applied a bead of silicone caulking between every sheet of plywood on the foundation.

We attached the first floor joists to the top of the foundation so the floor structure would resist the force of the backfill against the walls of the foundation. To make an especially strong connection at the back of our house, where the fill is deepest, we nailed an extra top plate and a 2x3 thrust plate onto the back foundation wall, and notched the joists to fit. To stiffen the end walls,

where the fill is deep, we added blocks between the two outer joists at 4-ft. intervals. For additional strength we glued (as well as nailed) the floor decks onto the joists.

It took us six days to erect the foundation walls, sheath their lower halves, attach the first floor joists and deck the first floor. We have read that experienced crews working with a small crane and prefabricated wall sections can completely erect AWWF walls in a few hours.

Finishing up—The concrete slab was poured after the first floor deck was completed. We prepared the floor by re-leveling the gravel in the foundation. This did not require as much accuracy as leveling for the footings—half an inch tolerance was acceptable. We shoveled excess gravel outside and laid a 6-mil polyethylene sheet on the gravel and a few inches up the walls as a moisture barrier. We also nailed screed boards of 1x3 treated wood around the sump basin and along the long sides of the two floor sections (one covering the front two-thirds of the foundation, and the other over the back third). Besides providing guidance in spreading the wet concrete, the screed boards helped hold the plastic in position.

The concrete truck pulled up to the front of our house, and the chute was put through the large window openings. We used a homemade chute extension, built from plywood and 2x10s, to reach the back third of the house. Aside from the person who delivered the concrete, we had one other to help us with the pour. The resulting slab was roughly 4 in. thick, with its surface about 1 in. above the bottom of the exterior studs. The openings between the studs provided plenty of air circulation so we used a gasoline-powered trowel for surface finishing.

After the slab was in, we finished the foundation sheathing. The excavator completed the drainage-pipe loop around the back of the house. The pipe was set lower than the footing plates all the way around. As in the bottom of the frost wall, it was surrounded with crushed stone. We had a cleanout installed in the highest section of the loop in case we ever need to flush the drainage system.

Next, we draped 6-mil polyethylene sheeting around the foundation from finished grade to just below the bottom of the footing plate, and protected its top with a 1x4 strip of treated wood caulked along its top edge. Gravel was used as backfill close to the foundation, and the finished grading included sloping all the surfaces away from the house to direct surface runoff away from the foundation.

The basement in the house we built has turned out to be very warm and dry—a very comfortable living space. The ease and speed of building the AWWF was outstanding, and the cost savings over the other types of foundations was significant. Our experience makes us wonder how long it will be until All Weather Wood Foundations displace concrete foundations, just as concrete foundations have displaced those of fieldstone. □

Irwin and Diane Post are forest engineers. They live in Barnard, Vt.

Capping a Foundation
One man's method for raising wood sills built too close to the ground

by Roger Allen

Many older homes were constructed with wooden sill plates too close to the damp ground. The primary reason for capping a foundation is to correct this situation by raising the sill. While this isn't the sort of task most homeowners would consider doing themselves, it is not as difficult as one might think. If the ground around the house can't be lowered, then with proper planning and careful workmanship, it's usually possible to raise the foundation.

In capping, siding is removed to expose the studs, and while the house is supported with temporary beams or house jacks, the studs are shortened to accommodate the cap. Then a new foundation is poured on top of (and in some cases, around) an existing foundation (drawing facing page, top.) Minimum clearances between sill and grade allowed by building codes vary, but 6 in. to 8 in. is the general rule. Reinforcing steel (rebar) and anchor bolts are included as in any other foundation. The new anchor bolts are an additional benefit of capping since many older homes were constructed without them.

The type of reinforcement used depends on whether the existing foundation is brick or concrete. In capping a concrete foundation, horizontal rebar and vertical dowels are added into the existing foundation, as shown on the facing page, top right. The size, amount and placement of this steel should be determined by an engineer. To cap a brick foundation, a saddle cap is constructed to strengthen the brick. This method requires the cap to encompass the brick foundation with a minimum of 3 in. of concrete on the sides. The height is determined by the necessary clearance above grade. Horizontal rebar is added to each side and to the top. In high caps a second horizontal piece may be required on top. A saddle tie, which is a piece of bent reinforcing steel, joins the horizontal pieces. The saddle tie is placed at approximately the same distance required of anchor bolts (generally every 4 ft.).

Capping a brick foundation has additional benefits. Old bricks and mortar tend to be more absorbent than new concrete, and capping prevents the bricks from acting as a wick, absorbing moisture from the ground and transferring it to the framing. It also adds reinforcing steel where there was none, and anchors the house to a solid wall of concrete.

Getting the house off its foundation—Moving the structure poses obvious problems. How do you pour concrete on top of a concrete wall that supports the house? How do you support the house when necessary framing is removed? And how do you pour concrete into that dark, underground world of spiders and snails, where a human can barely crawl, let alone work?

If the challenge of working under such conditions does not entice you, think of the money you will be saving. To cap 120 ft. of a foundation with an 8-in. wide and 12-in. high cap, you'll need about 3 cu. yd. of concrete. At a cost of $55 per cu. yd., that means $165 for concrete. You may need $100 more for a concrete pump, and another $100 for reinforcing steel, anchor bolts and miscellaneous hardware. Forming lumber could run yet another $100, and one or two helpers on the day of the pour perhaps $100 more. If you consult an engineer, it could cost you another $100. All this adds up to $665. I have seen estimates as high as $8,000 for the same job. Perhaps spiders aren't so scary.

Because the existing sill must be removed where the foundation is to be capped, temporary supports must be constructed. It is wise to cap a foundation in several sections and avoid supporting the entire house at once. I generally study how the house is normally supported and duplicate this support as closely as possible. This is one case when you should always overbuild your bracing. If you have any doubt as to how strong the temporary supports should be, consult an engineer or an experienced builder.

Support procedures—Where the floor joists run perpendicular to the foundation, a beam (usually a 4x8) can be run underneath them a few feet in from the foundation (figure 1A). The posts (one post every 8 ft. is usually enough) should be set on a solid pad in undisturbed soil. The pads must be large enough to distribute the load over the ground without sinking. On very solid earth, precast concrete piers make excellent pads. Heavy blocks of wood or timbers will often suffice, but in some cases I have had to dig a hole 4 in. deep by about 18 in. square to pour a concrete pad.

House jacks and shim shingles can simplify the installation of temporary supports. When the old framing is removed, the house will settle a little. To minimize settling, I aim for a snug fit of the temporary support posts. House jacks can be used in place of or in conjunction with the posts to snug the beam up to the floor joists. Shim shingles tapped between the posts and beam also help to ensure a tight fit. Support posts should be plumb and checked periodically for leaning. Be sure that all the posts and beams are secured strongly enough with diagonal bracing to withstand collisions from workers.

When the joists run parallel to the foundation, another method of support is called for. In this case a rim joist will be carrying the load when the studs beneath are removed or shortened. If the rim joist is not doubled, it should be at this time; to prevent tilting, add blocking to the adjacent joist (figure 1B). The distance between the temporary support posts will be determined by the allowable span of the joist. If the joist is doubled and nailed securely, it will increase the allowable span to about 6 ft. in a standard platform-framed, two-story house. When the house is supported this way, repairs must be made in sections between the supports.

If you want to repair the whole wall at one time, place 4x8 beams every 6 ft. perpendicular to the rim joist and support them at either end, as in the drawing on the facing page, bottom right. This technique allows longer spans of new sill to be installed at one time, but you have to remove more exterior siding than may be necessary, and you have to work around twice as much support timber. Nevertheless, it may be the best or the only technique for a particular job. Don't feel limited to only one technique for supporting a house. Many times a combination works well.

Removing existing sills—Once the house is supported, cut the existing studs to the proper height to accommodate the raised foundation. It is often easier to remove the exterior siding and work from the outside, but before choosing this approach, consider that exterior siding left in place makes an excellent concrete form; stucco also works well. On a concrete foundation, if there is adequate work space on the inside, leave the siding on and remove what is necessary after pouring the concrete. On a brick foundation, this consideration is irrelevant because the existing foundation must be encompassed on both sides.

The old studs must be cut off very straight. When the siding is left, a Sawzall is generally the tool to use, but making a straight and square cut with a Sawzall is difficult. Skilsaws cut square but in the cramped quarters you may be working in, Skilsaws are unwieldy, often dangerously so. A good sharp handsaw (and some elbow grease) is often the best choice.

Before cutting the studs, connect them with a well-nailed tie brace to keep the studs from moving once they lose the connection with the old

From *Fine Homebuilding* magazine (August 1981) 4:34-37

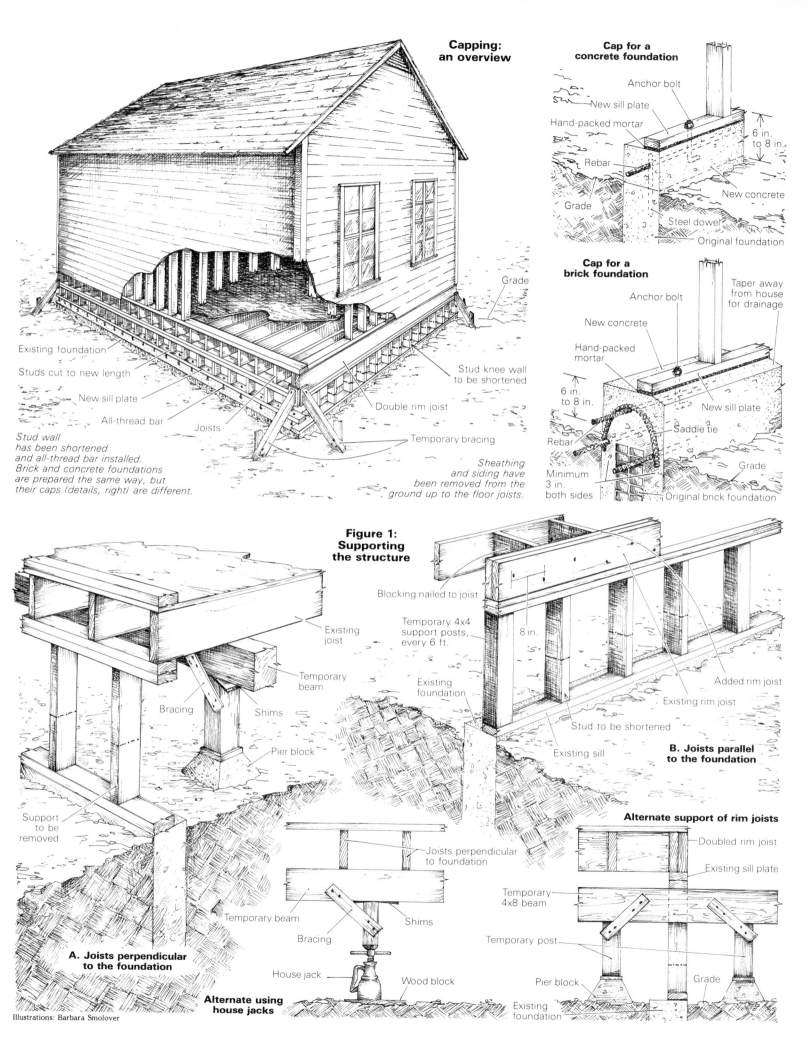

**Capping:
an overview**

Existing foundation

Studs cut to new length

New sill plate

All-thread bar

Joists

*Stud wall
has been shortened
and all-thread bar installed.
Brick and concrete foundations
are prepared the same way, but
their caps (details, right) are different.*

Grade

Stud knee wall
to be shortened

Double rim joist

Temporary bracing

*Sheathing
and siding have
been removed from the
ground up to the floor joists.*

**Cap for a
concrete foundation**

Anchor bolt

New sill plate

Hand-packed mortar

Rebar

Grade

New concrete

6 in.
to 8 in.

Steel dowel

Original foundation

**Cap for a
brick foundation**

Anchor bolt

New concrete

Hand-packed
mortar

6 in.
to 8 in.

Rebar

Minimum
3 in.
both sides

Taper away
from house
for drainage

Saddle tie

New sill plate

Grade

Original brick foundation

**Figure 1:
Supporting
the structure**

Blocking nailed to joist

Temporary 4x4
support posts,
every 6 ft.

8 in.

Existing
foundation

Existing
joist

Temporary
beam

Bracing

Shims

Pier block

Added rim joist

Existing rim joist

Stud to be shortened

Existing sill

**B. Joists parallel
to the foundation**

Support
to be
removed

**A. Joists perpendicular
to the foundation**

Temporary beam

**Alternate using
house jacks**

Joists perpendicular
to foundation

Bracing

Shims

House jack

Wood block

Alternate support of rim joists

Doubled rim joist

Existing sill plate

Temporary
4x8 beam

Temporary post

Pier block

Grade

Existing
foundation

Illustrations: Barbara Smolover

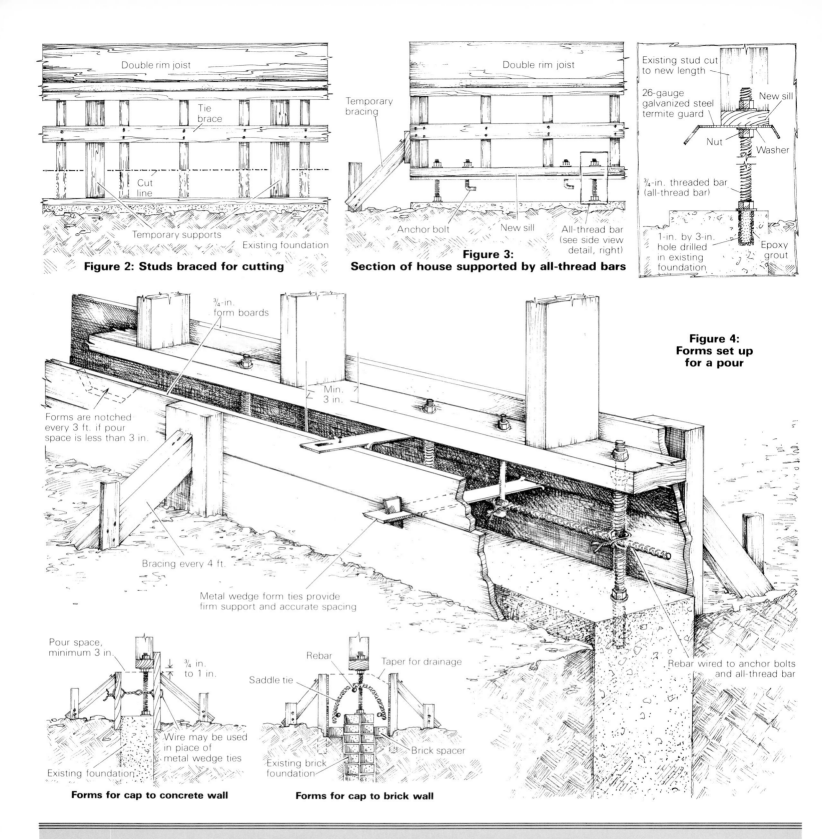

Figure 2: Studs braced for cutting

Double rim joist

Tie brace

Cut line

Temporary supports

Existing foundation

**Figure 3:
Section of house supported by all-thread bars**

Double rim joist

Temporary bracing

Anchor bolt

New sill

All-thread bar (see side view detail, right)

Existing stud cut to new length

26-gauge galvanized steel termite guard

New sill

Nut

Washer

¾-in. threaded bar (all-thread bar)

1-in. by 3-in. hole drilled in existing foundation

Epoxy grout

**Figure 4:
Forms set up for a pour**

¾-in. form boards

Min. 3 in.

Forms are notched every 3 ft. if pour space is less than 3 in.

Bracing every 4 ft.

Metal wedge form ties provide firm support and accurate spacing

Rebar wired to anchor bolts and all-thread bar

Pour space, minimum 3 in.

¾ in. to 1 in.

Wire may be used in place of metal wedge ties

Existing foundation

Forms for cap to concrete wall

Rebar

Taper for drainage

Saddle tie

Brick spacer

Existing brick foundation

Forms for cap to brick wall

Foundation capping: a step-by-step summary

1. Note the presence of water lines, sewer lines, meter boxes, wiring, and other obstacles that may affect the placement of temporary and permanent supports.
2. Support the house where capping is to be undertaken.
3. Nail a tie brace to studs above cut line.
4. Remove the old sill so your sawblade won't bind when cutting the old studs.
5. Cut old studs to proper height.

6. Install anchor bolts and all-thread bar to a pressure-treated sill.
7. Nail new sill to bottom of cut studs.
8. Drill holes in old foundation for all-thread and rebar if needed. (This may precede step 7 if workspace is limited.)
9. Grout all-thread into old foundation and tighten the nuts on the all-thread between the old foundation and the new sill. Remove temporary supports.

10. Hang rebar.
11. Build forms.
12. Pour concrete.
13. Vibrate.
14. Allow concrete to cure.
15. Hand pack mortar under new sill.
16. Remove forms.
17. Tighten anchor bolts.
18. Remove temporary bracing.
19. Replace siding if it has been removed.

sill. This step, shown in figure 2 on the facing page, is especially important where the siding has been removed.

Installing a new sill—Once the studs have been cut and the lower parts removed, nail the new sill to their bottoms. Select a good straight pressure-treated sill. When the sill is cut to expose a cross section, the lumber should be green all the way through. If you see blond lumber, the pressure treatment was not thorough. Use only well-treated sills—you won't want to do this job a second time.

It may be easier to drill the holes for the anchor bolts before the new sill is nailed to the bottom of the studs. Take care not to place a bolt where a stud will fall. The size and spacing of anchor bolts depend on local requirements. In areas prone to earthquakes, anchor bolts are commonly placed 4 ft. apart, and no more than 1 ft. from corners, ends, or unions in the sill. Elsewhere, 6-ft. spacing is the norm. Common anchor bolts range in size from $\frac{1}{2}$ in. to $\frac{5}{8}$ in. by 8 in. to 10 in.; $\frac{1}{2}$ in. by 10 in. is typical.

If termite-proofing sheet metal is to be added, fasten it tightly to the bottom of the sill before drilling and nailing. Drill through the sheet metal first using a hole saw attachment. Yes, this does wear the sawblades out fast.

If large sections of foundation are to be capped in one pour, all-thread bar should be installed. All-thread bar, or threaded steel rod, is metal stock threaded its entire length. It is available in various sizes from hardware stores or concrete accessory companies. At the same intervals that one would place a support post, the all-thread bar should be run from a 1-in. diameter hole drilled 3 in. deep into the old foundation through the top of the new sill, as shown in figure 3. Pack grout or an epoxy equivalent into the gap between the $\frac{3}{4}$-in. rod and the larger hole to ensure a firm connection with the old concrete. Place nuts and heavy washers at the bottom of the bar on top of the old foundation and directly under the new sill. When the nuts are tightened, the all-thread will secure the nailed sill firmly to the framing above and prevent it from sagging. When they are properly installed, the all-thread bars support the house and the temporary posts can be removed.

All-thread bar must be installed plumb or it will not carry the house properly. Although the all-thread bar is adequate to support the house in short sections, it does not supply any shear strength; it only carries a load from directly above. It is also necessary to attach temporary diagonal bracing to the corners of the house, as shown in figure 3.

If there is enough room above the new sill, the all-thread can be added after the sill is nailed in place. Otherwise insert it into the sill before installation. Holes in the softer brick foundations may be drilled with masonry bits and a normal electric drill. For a concrete foundation, considering that many holes have to be drilled, you can rent a Roto-hammer—a combination jackhammer and drill—from most equipment yards. Remember to drill the holes in the sill and foundation $\frac{1}{4}$ in. larger than the bolts to allow some leeway for installation.

Forming the cap—When the new sill is nailed to the bottom of the shortened studs and the all-thread is in place and tightened, it is time to install the rebar. The all-thread is a handy place from which to hang it. If the wall is to be poured in sections, the steel should extend 24 in. beyond to provide an effective tie to the next section when it is poured. If vertical steel is used, it should be inserted into holes at least 3 in. deep in the old foundation and grouted.

Because not much concrete is usually involved in pouring a cap, forming is relatively simple. Often all you have to do is to brace boards that are nailed in place to wooden or metal stakes. If the exterior siding is left in place as a form, make sure that weak points, such as the union of two pieces of siding or a crack in the stucco, are strong enough. If not, beef them up.

When the cap is to lie only on top of the old foundation and not encompass it, the outside form will rest tightly against the old foundation's side. If the siding has been removed, these form boards can be nailed with duplex nails to the framing above. Duplex nails should be used on all forms to allow for easy removal. The inside forms will also rest against the old foundation and require wood or metal stakes and form ties or wire for proper support (figure 4).

When the cap rests only on top of the foundation, leave space at the top of the inside form for pouring the concrete. Usually the size of the sill in relation to the size of the wall will allow for this space. If more space is required (3 in. is the minimum) notch the form at 3-ft. intervals at the top and build a funnel for the concrete to fall into the notch. If a termite guard has been added, you can temporarily bend the edge upward and out of the way before pouring the concrete.

A cap for a brick foundation requires a minimum 3-in. spacer every 4 ft. at the bottom of the form. Bricks work well for this purpose. After the concrete is in place and has been vibrated, use a small trowel to taper the exterior top away from the house for drainage.

Many people assume that because a cap will hardly be seen there is little reason to make it look good from within. Consequently, there is a tendency to build sloppy-looking forms on the interior side. But let me caution you: Sloppy-looking forms are often weak. Straight, neat forms are much easier to brace properly.

Pouring concrete—Most of the hard work is now over and the most exciting part is about to begin: the pour itself.

For small sections of foundation work, sacks of concrete can be mixed in a wheelbarrow and distributed via shovel or bucket. For larger amounts, a ready-mix truck should deliver your calculated amount. Always add at least 10% to the estimated load to allow for the inevitable tipped wheelbarrow. If the volume of concrete is very large or the workspace too awkward to drag buckets of concrete, you'll have to hire a concrete pump. Hire a grout pump and use concrete made with pea gravel, a concrete mix with aggregate no larger than $\frac{3}{8}$ in. The mix costs about $10 more per yd., but it allows you to use a smaller and less expensive concrete pump—the grout pump. A grout pump has the added advan-

tage of smaller hoses, which means less weight to drag around under the house.

Concrete pumps and concrete are usually ordered from separate companies—plan ahead to synchronize the two. Even so, on the day of the pour, one is likely to be late. Don't panic, it always happens. It is far better, however, to have the pump arrive early so it can be set up before the concrete arrives.

Before you begin to pour, be sure that the operator of the pump understands your instructions as to starting and stopping the pump. Learn early what the delay between your instruction and his response will be. If you are far under the house, another person can relay your commands. I can recall several times I was almost buried in overflowing concrete because a pump operator was talking to a neighbor or eating his lunch. If you have calculated the volume of concrete as closely as you should have, you don't want it wasted.

A critical aspect of pouring is to make sure that no spaces are left unfilled in the forms. The concrete must be vibrated so it settles everywhere. Vibrating guards against the weakening effect of a honeycombed wall. There are several methods of vibrating. An electrical vibrator, designed specifically to be placed into the concrete on large pours, is rarely necessary for capping. One good technique for capping is to tap the sides of the forms with a hammer, as well as packing the concrete from the top with a stick. Another way is to remove the blade from a Sawzall and place the shoe of the Sawzall against the outside of the form; when running, the Sawzall acts as an excellent vibrator.

Vibrating increases the stress that the poured concrete puts on the forms, so the forms should be watched as carefully when vibrating as they were during the pour. If the bracing seems inadequate, stop pouring or vibrating and shore up the weak sections immediately.

Because concrete shrinks as it dries, a small gap would be left under the new sill if the concrete were poured all the way to the bottom of the sill and then allowed to dry. To solve this problem, leave a $\frac{3}{4}$-in. to 1-in. gap between the newly poured concrete and the bottom of the new sill. After the concrete has cured, you can hand-pack mortar into this space. If the gap were smaller, filling would be more difficult.

All concrete spilled onto the forms and stakes should be cleaned off before it hardens, making it easier to strip the forms. Many a worker has cursed his lack of foresight while struggling to pull stakes or to pry boards that are embedded in hardened concrete. You can pull off the forms the day after the pour, but I usually leave the temporary supports in place for a couple of days to allow the concrete to cure and attain its full strength before accepting the weight of the house. After the concrete has cured and the hand-packed mortar has hardened, the anchor bolts should be tightened. If siding was left on the outside as a form, it should now be raised to the proper level above grade. Your house now sits high and dry. □

Roger Allen is a general contractor in the San Francisco area.

Floor Framing

With production techniques and the right materials,
a solid, squeakless floor is a day's work

by Don Dunkley

Rough framing isn't measured in thirty-seconds of an inch, and for good reason. Although a shabby frame can put finish carpenters in a murderous mood, it gets covered up and forgotten behind siding, drywall and paneling. But floors, if done poorly, will come back to haunt you. That squeak just outside the bedroom door is an annoyance for which there is no quick fix. But the problems can go beyond the floor itself. The blame for eaves that look like a roller coaster once the gutters are hung often rests squarely on a carelessly built floor two stories below.

Haste doesn't usually create these problems; using inappropriate materials and not knowing where to spend extra time does. In fact, by using production techniques and materials, rolling second-story joists (tipping them up and nailing them in place) and sheathing them with plywood on a modest-sized house is a day's work for my partner and me.

What a floor does—Most builders are very aware when they hang a door of the abuse it will get, but they don't think in the same way about the floor they are framing. This skin is the main horizontal plane the building relies on to transfer live loads (ones that are subject to change, like furniture and people) and dead loads (primarily the weight of the structure itself) to the bearing walls, beams and foundation. Critical here is the amount of deflection in both the sheathing and the joists. The subfloor must be stiff enough to handle both general and concentrated loading, and to support the finish floor that rests on it without too much movement. But at the same time it must be flexible enough for comfortable walking and standing, something a concrete slab can never be.

Floor systems—There are many different ways to build a floor, and the preference for one system over another has a strong regional flavor. Floor trusses—either metal or wood—are becoming popular because of their prefab economy, and because they can eliminate the need for bearing walls, beams and posts by free-spanning long distances. Girder systems are still used over crawl spaces in some areas. They are generally laid 4 ft. o. c. with their ends resting on the mudsill of the foundation wall, or in pockets cast into the wall itself, or on metal hangers attached to the foundation. The interior spans of the girders are typically posted down to concrete piers. Girders are decked with either 2x T&G or very heavy plywood designed for this purpose.

But most wood-frame houses still use joists. And whether a floor rests on foundation walls on the first story or on stud walls on a higher level, its elements are pretty much the same. Joists are typically 2x lumber, laid out on 12-in., 16-in. or 24-in. centers and nailed on edge. They are held in place on their ends by toenails to the sill or plate below, and then attached to a perimeter joist, or blocked in between. Blocking or bridging has also been used traditionally to stabilize a floor at unsup-

ported midspans. (For more on this, see the next page.)

The last element of any joist system is the skin—usually plywood sheathing. This decking not only forms the continuous horizontal surface, but is also the key to the integrity and structural continuity of any wood floor. I'll talk more about the choice of material and how to apply it later.

In the simplest floor system, joists rest on the mudsill of exterior foundation walls, or on the double top plate of the stud walls on higher stories. With the long spans found in most floor plans, joists have to be either very deep (say, 12 in. or 14 in.), or supported at or near the center of their length by beams or bearing walls, creating two shorter spans. In the case of even greater building widths, joists can come from either side of the building, lapping each other over these supports, with blocking nailed in between.

In much of the country where basements are used to get below the frost line, the joists bear on or lap over built-up beams of 2xs spiked together. These are usually supported by Lally columns—concrete-filled tubular steel posts with flanges top and bottom—or by hefty wood posts.

In the West, where I live, the first-story floor usually stretches over a crawl space. Perimeter support for the joists is provided either by stemwalls—mudsill-capped low foundation walls (most crawl spaces use the minimum allowable height of 18 in. from grade to the bottom of the joists)—or by cripple walls (sometimes called pony walls) that are framed up on hillside foundations to reach the first-floor level.

For supporting the joists in mid-length or for lapping them, floors over crawl spaces either use a carrying beam, or girder, on 4x posts that rest on concrete pier blocks, or use a crib wall—a low, framed wall studded up from an interior concrete footing that runs the length of the building.

Unlike girders, which are often crowned and leave a hump in the finish floor, crib walls can be plumbed and lined to make a perfectly straight, level surface. They will also shrink less than a large girder, and just feel a lot more solid to me. The posts under a girder are toenailed at the bottom to the wooden block that caps each concrete pier. This block inevitably splits, making the whole post-and-beam assembly feel a little cobbled together. I've also learned that concrete contractors get bored with straight lines at the point when they begin laying out piers, so supporting posts often get precious little bearing.

Bolting down the sill—Floors need to be level, square and precise in their dimensions. On a second floor, these conditions will depend on your accurately plumbing and lining the first-floor walls; but on the first floor, setting the mudsill is the critical step. First, check the perimeter foundation walls with a steel tape for the dimensions shown on the plans. Then pull diagonals to check for square. Run 3-4-5 checks on the corners for

square if the floor isn't a simple rectangle. Once you're sure of your measurements, snap chalklines that represent the mudsill's inside edge on the top of the foundation walls.

Be thinking ahead to the sheathing (called shear panel where I build) and siding if the foundation walls aren't arrow-straight. If these are one and the same, the plywood will need to hang down over the concrete foundation wall an inch or two. This means that you're better off having the mudsill overhang the foundation slightly, rather than the other way around. If finish siding is used over sheathing, this isn't as important, since the sheathing can butt the top of the foundation where it bows out from the mudsill.

Place the mudsill (either pressure-treated or foundation-grade redwood) around the perimeter of the foundation, and cut it to the lengths needed to fit on the stemwalls. You'll find foundation bolts placed on maximum 6-ft. centers; code requires each piece of sill to be held by at least two bolts. The cut sills should be laid in position on their edges along the top of the stemwalls so that they can be marked for where the anchor bolts fall along their length.

After carefully squaring these marks across the face of the sill, lay them flat so you can mark the bolt positions across their width. Measure the distance from the chalkline on the top of the stemwalls to the center of each bolt. Then transfer that measurement to the face of the mudsill, taking care to measure from the inside edge of the mudsill each time.

Production carpenters who do a lot of layout use a bolt marker. One of these can be made from an old combination-square blade, as shown in the drawing below. The end of the metal blade is notched so that it will index from the centerline of a ½-in. foundation bolt. At 5½ in. from this end for a 2x6 mudsill (or 3½ in. for a 2x4), a hole is drilled in the blade and fitted with a screw or nail.

To use this marking gauge, the mudsill has

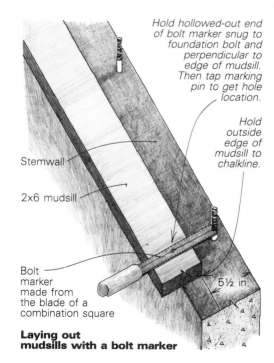

Hold hollowed-out end of bolt marker snug to foundation bolt and perpendicular to edge of mudsill. Then tap marking pin to get hole location.

Hold outside edge of mudsill to chalkline.

Stemwall

2x6 mudsill

Bolt marker made from the blade of a combination square

5½ in.

Laying out mudsills with a bolt marker

Bridging midspans

Although the days when all carpenters learned the trade through a formal apprenticeship are gone, construction knowledge is still passed down through the ranks, from those who know to those who want to. This kind of conservatism is often valuable, but it also means that outmoded methods die hard. Bridging between joists at midspan is a good example. Even on relatively short spans, a lot of knowledgeable builders are convinced that bridging guarantees a stiffer floor. But convincing studies show that it doesn't, and that information isn't new.

Bridging is a continuous line of bracing that runs perpendicular to the direction of the joists, and is installed between them. It can take the form of cross bridging (diagonal pairs of wood or metal braces that form an X in the joist space) or horizontal bridging (continuous, full-depth solid blocking).

A study done by the National Association of Home Builders Research Institute Laboratory in 1961 put most of the myths to rest. It proved with laboratory and field tests that bridging at midspan had little effect in stiffening a floor once it was sheathed. It was also of little help in transferring concentrated loads laterally to other joists, reducing floor vibration or preventing joists from warping. Bridging actually does all of those things as you install it,

but once the plywood subfloor is on, its net effect is negligible.

Still the myth persists. Building in the 1970s, I put in midspan solid blocking every 8 ft. on floors I built, to meet the code and because I was concerned with quality building. But there is only one way in which midspan bridging improves a floor substantially, and that is in providing lateral stability. Because joists are on edge, they tend to tip unless they are pinned in at their ends and at support points like girders and bearing walls. Plywood sheathing nailed and glued on top also helps. But over long spans, the joists can use a bit more help.

Of the three model codes in the U. S., only ICBO's Uniform Building Code requires using cross bridging or solid blocking with plywood sheathing under normal conditions, and that is only when the depth-to-thickness ratio is 6 or more, as with 2x12s. It is then required every 8 ft. along the span of the joists. BOCA requires bridging or blocking only if the joist depth exceeds 12 in. under normal loading conditions, and SBCC allows you to skip it entirely in single-family residential buildings.

If you do need to install midspan bridging or blocking on a floor you're building, a few simple tricks will make it go faster and eliminate some of the squeaks. First, you must decide between diagonal bridging, which can either be wood 1x3s or the newer sheet-metal bracing, and solid blocking.

I think that cutting and nailing the old wooden cross bridging is needlessly time-consuming, although all the standard construction texts cover it in detail. If I were going to use bridging, I would choose the metal variety that doesn't use nails. These straps have sharp prongs on their ends, and eliminate the squeaking you get from nails rubbing on the metal bridging when the joists flex underfoot. This kind of bridging can be installed from above or below the joists; a combination of the two is even better. Drive the upper ends of the bridging into the top of each joist face before sheathing, and return to drive the bottoms in from below when the joists have done most of their shrinking, and initial loading has flattened them some. Make sure to separate each member of the pairs to prevent contact squeaking.

Solid blocking has to be done just before sheathing, but it's straightforward. If you're going to insulate the floor yourself, it's the best choice. To begin, chalk a line across the joists where the blocking goes. This is typically at the center of any span under 16 ft., or in two rows equally spaced if the span is greater. Alternate the blocks on either side of this line so you can face-nail them through the joists.

Set a plank on top of the joists, parallel to the chalkline and about a foot away. Set it to the left of the line if you're right-handed. Spread out your blocks along this plank. Drive three 16d nails partway into the face of each joist down the line, alternating between the right and left side of the line. This allows you to position each block and then anchor it with a single blow of the hammer. Once you've traveled the entire length of the joists nailing the blocks on one end, you can turn around and travel back the way you came, nailing the other ends.

Construction adhesive recommended for solid blocking is supposed to cut down on squeaks, but it's messy and time-consuming. It's more practical to drive shims in any gaps that open up between the blocks and the joists. Don't use vinyl sinkers (vinyl-coated nails) or brights (uncoated steel nails). These will loosen up and squeak much sooner than ring-shanks, screw nails or hot-dipped galvanized.   —*Paul Spring*

to be held flat with its outside edge on the chalkline. This is most easily done from inside the foundation. Make sure that the sill is exactly where you want it along its length. Then by holding the bolt marker perpendicular to the length of the sill, with the notch of the gauge pressing against the bolt, tap on the marking pin with your hammer. Whatever the location of the bolt, the correct bore center will be left on the sill.

After the sills are laid out for bolts and drilled with a slightly oversize bit (a 9/16-in. bit for 1/2-in. bolts is a good compromise between requirements for a snug fit and giving the carpenter a break), place them on the stemwall and hand-tighten the nuts. If termite shields or sill sealers (wide strips of compressible insulation that act as a gasket to keep air from passing between foundation and framing) are to be used, be sure to lay them in underneath the mudsill. Now is also the time to adjust the foundation for level by shimming under the mudsill where necessary with grout. Go back and cinch the nuts down tight once everything is set.

With a hillside foundation where the joists will rest on a cripple or pony wall, adjustments for level are made in the cripple studs themselves rather than by shimming the plate. If the foundation is stepped down the hill, these cripple studs will be in groups of various lengths. If a grade beam that parallels the slope is used, then the cripples will have to be cut like gable-end studs. In this case the beveled end will be nailed to the mudsill and blocked in between. In either case, getting the top plates precisely level (and straight) is worth the trouble. Do this by shooting the tops of the studs with a transit, or by setting up a string-and-batterboard system. Using a spirit level for long runs just isn't accurate enough, and trusting a hillside foundation to be true is a mistake you'll make only once.

Joist layout—The first thing I do to lay out the sills for rolling joists is to get out the plans again and find all the exceptions to the standard layout. These include stair openings that have to be headed off, lowered joists for a thick tile-on-mortar finished floor, plumbing runs that need to stub up on a joist layout, and the double joists used for extra support. Even a simple floor will require doublers to pick up the weight of the building at its most concentrated points.

The first place to double is at each end of the foundation, parallel to the direction the joists run. When joisting a second story, sandwich short 2x spacer blocks between two joists for a tripler (drawing, facing page). This will give you backing for the ceiling drywall below. Locate all interior walls running parallel to the joists and lay out the sills (and carrying beams if you have them) for doublers too. Since some plumbers feel that joists have just as much integrity cut in half as they do whole, it's a good idea to have a chat with your pipe bender before you complete the joist layout. This will save a lot of headaches later. For instance, the doublers under plumb-

Heading off the joists to make room for DWV lines, as shown here, is a lot easier if plumbing runs are considered before framing begins.

ing walls will have to be spread for waste lines that run parallel to the joists, or headed off with solid blocking to accommodate perpendicular plumbing runs (photo above). Doubling joists under a heavy cast-iron bathtub is also a good idea.

Next, lay out the joist spacing shown on your plans. This layout has to be adjusted so that the butting edges of the decking will fall in the center of a joist, allowing two sheets to join over a single joist. Stretch your tape from

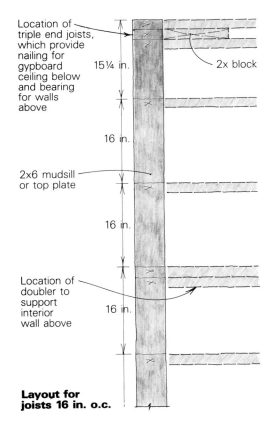

Location of triple end joists, which provide nailing for gypboard ceiling below and bearing for walls above

15¼ in.

2x block

16 in.

2x6 mudsill or top plate

16 in.

Location of doubler to support interior wall above

16 in.

Layout for joists 16 in. o.c.

the outside edge of the sill and make a mark ¾ in. (or half the thickness of a 2x joist) shy of the joist-spacing dimension—15¼ for 16 in. o. c. Put your X on the leading side (far side) of this mark for the first joist. The rest of the layout can be taken from this first joist at the required centers without any further adjustment. If your floor has joists coming from each side that lap over a beam or wall, the layout on one side will need to be offset from

the layout on the other perimeter sill or plate by 1½ in. to account for the lap.

After you have completed laying out the sills and midspan beams, check again to make sure that they jibe, and that the centers are accurate for the plywood. Now is the time to think; once you begin rolling joists, you just want to move.

Unlike rim joists, the end-blocking method requires cutting special-length blocks to correct for layout and accommodate double joists.

Getting ready to roll—There are two ways to secure the ends of the joists. The first is to run them the full width of the house, and block in between at their ends (photo above). This system has some structural advantages since the joist ends are locked in, but with all of those short, brittle pieces involved, it requires more fussing to get straight lines. The other way is to cut the joists 1½ in. short on each end, and run a 2x rim joist (also called a header joist, ribbon or band joist) perpendicular to the joists and face-nail it to them.

Whichever system you use, all the joists should be laid flat on the sills right next to their layout marks before you begin to frame. Stocking is basically a one-man job. As you pull each one up, sight it for crown. Many framers mark the crown with an arrow (make sure it points up), but it's not necessary if you start at one end of the building, and lay each joist down so that the crowned edge is leading. This way, when you return to the beginning point, you'll be able to reach for the leading edge of each one and tilt it up or roll it, leaving the crown up. Save the really straight lengths for rim joists. If any of the joists are badly crowned, cut them up for blocking.

Rim joists—The rim joist for one side of the building should be installed before any of the joists. Mark its top edge with the same layout as the sill it will sit on. If a large crown holds the rim up off the top plate or mudsill, cut the rim joist in half on a joist layout mark. Also, if this floor is over a crawl space, you will need to cut in vent openings. Place these every 8 ft., starting 2 ft. in from the corners, where air tends not to circulate. Typical vent openings are 5 in. high and 14½ in. long; these fit nicely between joists on 16-in. centers.

Install the rim joist with toenails from the outside face of the board down through the sill or top plate every 16 in. Once nailed, string the outside edge to make sure it is

straight and adjust it in and out if necessary with a few more toenails. Now snug up the joists that are resting flat on the plate or sill so they butt the inside face of the installed rim joist (this is easily done by catching the face of the joist that's up with the claw of your rip hammer and pulling toward the outside of the building, as shown below). If any joist ends are badly out of square, trim them.

Using the perimeter mudsills or plates, which have already been carefully plumbed and lined, as a kind of tape measure, mark

The author snugs a joist against the rim to check the end for square, using the building itself as a ruler for cutting the joists to length.

Kiln-dried framing makes better joists, but in the West green Douglas fir is much more common. Even so, the ideal moisture content of a joist is 15%. Many of the problems with floors—in particular squeaking and nail-pop—can be traced to wet joists shrinking away from nail shanks as the wood dries. Wet joists can permanently deflect once they begin to dry, causing a swale in the floor. Most of the so-called settling in a platform-framed house is due to the joists shrinking across the grain. Much of this can be eliminated by using dry lumber. Putting down a plastic moisture barrier over the earth in damp crawl spaces is a good way of keeping the floor above dry.

Cutting to length. The joists are squared up 1½ in. short of the building line (below) to leave room for the rim joist, and cut to length. With their crowns marked by an arrow, they are ready to be tipped up.

Rolling the joists. Joisting (right) is fast if a routine is established between partners. On the open side of the joists, an end toenail holds the joist on layout; toenailing the joist face to the plate will follow. On the other end, the rim joist is face-nailed to the joist with three 16d nails, then toenailed to the plate. The end joist in the foreground will be doubled later.

Installing the other rim. With joists nailed off, the open side gets its rim (bottom). This floor uses 2x14s; 2x10s are more typical.

each joist where it extends over the perimeter wall opposite the installed rim. Square up a line 1½ in. in from this outside mark, and cut. For a floor with more than one span, index off the outside of the girder or crib wall, and use the rule of thumb for lapped joists to determine if they will need to be cut. The joists are now ready for rolling.

With the rim joist installed, rolling goes fast, as the joists can be quickly tipped up and nailed in place—no special blocks to slow you down and no need to stretch a tape measure to make sure you're keeping the layout, as in end blocking. Rolling joists (photo facing page, top right) is a synchronized, almost fluid process that requires few words once partners are used to working together. With one of you on each side of the building, and beginning with the doublers at one end or the other, reach forward at the same time for the top of the joist, tip it up and place both ends of the joist on their layout marks. On the side of the building where the rim is already installed, drive three face nails through the rim into the end grain of the joist. If a joist is a little lower in height than the rim when tipped up, hold it up flush to the top edge and shim

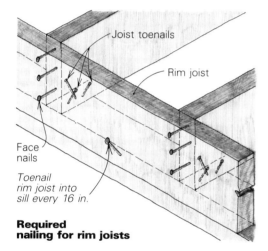

Required nailing for rim joists

Labels in figure: Joist toenails; Rim joist; Face nails; Toenail rim joist into sill every 16 in.

up after the joists are rolled. The other member of your crew on the opposite side will drive a single toenail through the end of the joist, down near the bottom, into the top plate or mudsill. Then, both of you drive two toenails through the leading face of the joist down into the plate or mudsill (the side facing the joists that haven't yet been tipped up), as shown above. Once you've rolled all the joists, you can reverse direction, and work your way back toenailing the other face of the joists to the plate or mudsill. This is called backnailing. Then the rim on the open side of the joists should be hauled into place, toenailed to the plate or sill, and face-nailed to each joist (photo facing page, bottom).

End-blocking method—If you choose end blocking over a rim-joist system, each joist will need a block, which should be precut and stacked on top of its joist as it lies flat and ready to be rolled. A radial-arm saw set up with a stop gauge is the easiest way to gang-cut blocks, but a skillsaw will do. For a 16-in.

layout, the blocks should measure 14⁷⁄₁₆ in. The more accurate your blocks, the fewer specials you'll have to cut while you're joisting. The first block, however, will be a special to account for the outside doubler and the ¾-in. adjustment that was made in the layout for the sheathing. After all blocks are cut, put them on the joists that are ready to be rolled. Proceed to roll the joists first by setting up the outside doubler. If you have to use more than one piece to make the length, then make sure that the butted splices are at least 4 ft. from each other for these double joists.

Once the doubler is in place, toenail the first special block to the doubler and to the sill or plate—first with a toenail through the

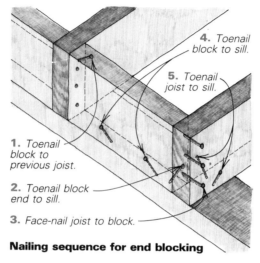

1. Toenail block to previous joist.
2. Toenail block end to sill.
3. Face-nail joist to block.
4. Toenail block to sill.
5. Toenail joist to sill.

Nailing sequence for end blocking

top of the joist into the block it butts against; second, with a low toenail into the sill or plate through the end grain of the block on the leading end. Then tip up the next joist in front of you and face-nail through it into the end grain of the block with three nails. Finally, drive two toenails from the outside of the block into the sill or plate (drawing, above) and toenail the joist itself down. Tip up all joists in the same manner. Although the plate is laid out, it is a good idea to measure joists every 4 ft. or so to make sure they aren't wandering off the layout. Shorten a block or cut a longer one to adjust if needed. This will take a lot of frustration out of the sheathing later on.

Lapped joists—In most floor plans, the distance to be spanned is long enough to require that a girder or crib wall be used to support the joists, or as a bearing surface where two runs can lap. Double-check the height of this girder or wall to avoid the typical center hump that a lot of floors develop after a year or so. To tell the truth, this support is best set slightly too low than too high. Crib walls will shrink less than most beams, but using dry lumber will help in both cases.

Joists can be butted and a plywood or lumber splice-plate at least 24 in. long used to join them, but most often they are lapped. The rule of thumb I've always used is that the joists should lap each other by at least 4 in., but that neither one should extend beyond the far edge of the girder or crib wall that sup-

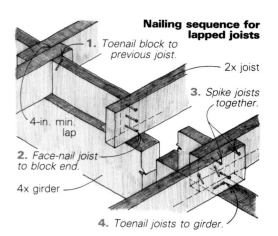

Nailing sequence for lapped joists

1. Toenail block to previous joist.
2. Face-nail joist to block end.
3. Spike joists together.
4. Toenail joists to girder.

Labels: 2x joist; 4-in. min. lap; 4x girder

ports it by more than this joist's nominal depth. The joists should be toenailed to the support beam or plate through each face, and then blocked solid in between, as shown above. These blocks have to be shorter than end blocks to make up for the extra joist. Pay attention to the layout by pulling a measurement from the end of the building every 4 ft. to make sure that the joists aren't bowed in or out from the layout on the perimeter, and that the subfloor joints will fall on a joist. (Remember, you may be dealing with two layouts. If the joists lap, the layout on one side of the floor will differ from the other by 1½ in.)

Schedule the rough plumbing after joisting is completed and before you begin the sheathing. If you are using forced-air perimeter floor heat, the ducts should be installed as well. Be sure to block around the heat registers in the floor to provide backing for the sheathing once the subs have finished the rough-in. Then call for inspection and get it signed off. This is also the time for insulation.

Sheathing—Joist systems can use either a single or double layer of sheathing. The traditional approach is to sheathe the joists first with subfloor—typically ½-in. or ⅝-in. square-edge plywood laid with its face grain (and long dimension) perpendicular to the joists. This forms a working platform for rough construction, but is covered over with underlayment (either plywood or dense particleboard) ranging from ⅜ in. to ¾ in. This produces a thick floor with little flex, and more resistance to sound transmission so the patter of little feet is muffled in rooms below.

However, most of the houses I build use a single-skin, combination subfloor and underlayment. It requires half the labor and produces a serviceable floor. These plywood panels are tongue-and-grooved on their long sides (the ends will always fall on a joist for continuous support), and are structurally rated to span the joist spacing by themselves. It's possible to order square-edge panels, but with these, continuous edge-blocking between joists is required. Single-skin panels range in thickness from ¹⁹⁄₃₂ in. to 1⅛ in. (known as 2-4-1). Always use the thickest you can afford. While code allows ⅝-in. plywood on 20-in. centers, and some builders are already calling for its use with glue on 24-in. centers; ¾-in. plywood on 16-in. centers has my vote. Inch-

Sheathing the floor. The first step in sheathing is to establish a course line 4 ft. in. from the rim with a snapline (top). Installing T&G plywood using a 2x4 beating block and a sledgehammer is a job for two. The author, in the rear of the photo above, is using both his shifting weight and his hammer to coax the tongue into the groove. This second course begins with a half sheet in order to stagger the joints. On this job, white glue was specified; construction adhesive is far superior.

and-an-eighth plywood (2-4-1) is even stiffer, and can be used on 16-in. or 24-in. centers even though it was designed for use over girders placed at 4 ft.

As in underlayment plywood, the voids in the face veneers and the plies directly beneath them in combination subfloor and underlayment are plugged so that the wood resists point loading (consider what high heels can do to plywood with interior voids). The panels are also drum-sanded to a precise thickness. With this plywood, about the only place a second layer is used around here is with vinyl flooring in baths and kitchens. Underlayment is particularly important with thin flooring because it provides a new dense surface that hasn't taken the abuse of construction traffic. Installed just before the finish flooring, it lessens the likelihood of joint lines or nailheads broadcasting through flooring. This extra layer also brings the level of these finished floors flush, or nearly so, with areas of the house that may have padding and carpet, tile or hardwood flooring.

Setting plywood—Check T&G plywood when it's delivered. If either long side is banged up, crushed by banding, or looks like it's been sitting out in the rain, don't even let them unload it. The tongue is difficult enough to thread into the groove without sabotage. Once you've got the material on site, treat it with care. Cover it even if clear skies are predicted.

To sheathe the floor using plywood, first snap a line 4 ft. in from one side of the building, using a level to plumb up from the mudsill or top plate of the outside wall below for accuracy. The butt joints of plywood have to be staggered between courses (there should never be an intersection of four corners), so you should begin either the first or second course with a half sheet. Remember that you are laying the plywood down with its length running perpendicular to the direction the joists run. Begin the first course with the tongue sitting on the rim joist at the perimeter of the building. This will leave the groove as the leading edge. The next course of plywood will have to be threaded onto the previous course by driving the new sheets with a 2x4 block and a sledge hammer. The grooved edge will take this abuse; the tongue won't.

If you have lapped joists, the layout for the sheathing will be different from one side of the floor to the other by 1½ in. If the lap occurs at a natural break for the plywood, the next course can just be slid back 1½ in. so that it continues to butt on a joist. But if a plywood course covers the joist lap, then you'll have to spike 2x4 scabs to each joist to get bearing for the plywood at this transition.

Gluing down the plywood with an elastomeric construction adhesive in addition to nailing enables the plywood and joists to act as one integral unit, creating what amounts to a T-beam. Using glue will lessen creep and squeaking and increase floor stiffness. It also lets you cut down the nailing schedule on plywood ¾ in. thick or less from 6 in. on the edges and 10 in. in the field, to 12 in. for both

using 6d (although I prefer 8d) screw nails or ring-shank nails. Construction adhesive comes in tubes that are used with a caulking gun. Using large tubes saves time, and buying by the case will save you a few bucks. White glue is better than nothing, but it won't bond as well as the neoprene-base adhesives do.

Apply a bead of adhesive continuously on every joist, but don't work too far ahead of yourself because it will skin over rapidly. Flop the plywood down carefully so that it's very close to where it belongs. On the first course, aim for the chalkline, and then nail off the sheet with the exception of the last 6 in. of width along the groove. Nail this little bit once the next course is driven into place or you will pinch the groove and make threading the tongue nearly impossible.

Plywood associations and textbook writers agree that this kind of combination subfloor and underlayment should be spaced ⅛ in. at edges and ends. The climate where I build is very dry, and this practice isn't common, but I can see that you might need to allow for expansion in some climates. If you do use this recommended spacing, you will need to trim panels for length occasionally since you will be adding slightly to the layout each time. Or you can scab on a 2x4 to the existing joist to get more bearing. This technique is particularly useful at the end of a course of plywood when the floor is just slightly longer than the last full sheet will stretch. Although plywood should always span at least two joist spaces, the exception is where a small piece rests completely on solid joists such as a double or triple. If your layout falls just short of the end of the floor, you can add a scab to the end joists to support the last full piece of plywood, and fill in with a ripping.

After the first course of plywood is nailed off, you can begin the next course, remembering to stagger the joints. This is best done with three people. The first lays down the bead of adhesive ahead of the plywood and drops back occasionally to nail sheets that have been tacked before the adhesive sets up. The other two will be cutting and setting the plywood. The only trick to setting is a bit of coordination between the person handling the sledge on the leading edge of the new panel, and the person using his feet and body weight to thread the seam (photo bottom left).

Once set, each sheet should be nailed at all four corners and at each joist, which will help whoever is nailing off keep the nailing lines straight. Nail off the plywood with ring-shank or screw nails because they hold much better than conventional nails, particularly in slightly green joists. Make sure the nails go straight into the joists. Nails that break through the sides will cause squeaks. It's worth sending someone below to drive out any that miss.

Once all sheathing is nailed off, trim off overhanging edges where necessary with a circular saw. Your efforts will result in a good strong subfloor that will withstand all the wear and tear a household will give it. □

Don Dunkley lives in Cool, Calif.

Parallel-Chord Floor Trusses

Strong and efficient, floor trusses may someday replace
wood joists as the builder's favorite floor frame

by E. Kurt Albaugh

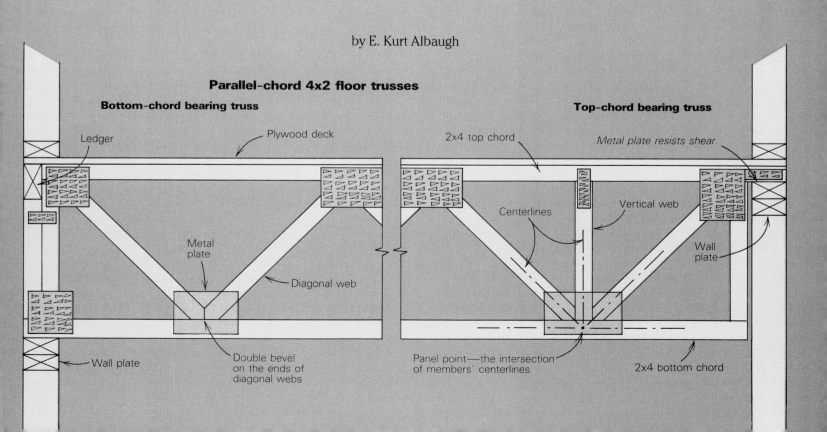

Parallel-chord 4x2 floor trusses

Bottom-chord bearing truss

Ledger

Plywood deck

Metal plate

Diagonal web

Wall plate

Double bevel
on the ends of
diagonal webs

Top-chord bearing truss

2x4 top chord

Metal plate resists shear.

Centerlines

Vertical web

Wall plate

Panel point—the intersection
of members' centerlines

2x4 bottom chord

Once associated more with commercial construction, structural trusses are becoming increasingly popular among home builders. Though most builders are familiar with roof trusses, fewer builders realize that floor trusses can be used quite effectively in residential construction, too. They offer a number of structural and economic advantages, and can easily be incorporated into the design of a home without significantly changing the way a builder builds.

Structural advantages of trusses—Traditionally, the size of rooms in a home has been based largely on the span limitations of standard wood joists. Floor trusses with the same depth as joists can be used over longer spans, and this means that rooms can be larger, with less space obstructed by columns or unnecessary partitions. Trusses have a much greater variety of depths than wood joists do, and therefore a much wider range of spans and strength.

Trusses rarely warp. With a joist floor, natural warpage in the members can lead to an uneven floor deck. This problem can be reduced somewhat by culling out the warped joists before installation, but that increases wood waste.

Another advantage of floor trusses is that they permit ducting, plumbing and electrical service to be run easily between the open webs. With a joist floor system, holes must often be cut through each joist in order to run electrical and plumbing lines.

Since floor trusses are sized by the fabricator, they're delivered to the job site in lengths that meet the specific requirements of the project. With joists, the material must often be cut on site to fit, which wastes wood and increases the handling of material. Installation of a truss is relatively easy, because the truss edge is 3½ in. wide. This makes it more stable while it rests on the wall plates, and easier to nail to.

A floor-truss system is engineered, while a joist system usually isn't. This means that greater floor loads can usually be carried, and deflections are better controlled. Engineering also makes for a more efficient use of material (strength vs. weight), resulting in a lower per-square-foot cost.

Cost advantages of trusses—The builders I've worked with who use floor trusses believe that a floor-truss system is cost effective. But exactly how much so is difficult to determine since labor and material costs vary greatly throughout the United States. As a general rule for the Houston area, material costs are about $.90 per sq. ft. of floor (with trusses 24 in. o. c.). If the job is a long way from the manufacturer, shipping can increase the costs significantly. A joist floor system, on the other hand, costs about $.85 per sq. ft. of floor. This means that material costs for a truss floor are about 6% higher than for a joist floor system. But when the time saved for installing a truss floor (as much as 40%) is taken into account, the truss floor turns out to be about 10% less expensive overall.

What are the time savings? Quicker framing means a faster construction schedule, with attendant savings on construction financing. And because plumbers can route the pipes through the open webs of the trusses, no time is wasted by having to bore holes through joists. The electrician and HVAC (heating, ventilating and air conditioning) installers save time as well. Exactly how much time each tradesperson saves is hard to determine, but it can add up over the course of a job.

On the production side, most wood-truss fabricators find that there is little difference in production costs between 12-in., 14-in. and 16-in. deep trusses, so these tend to be comparable in cost to the builder. But when trusses get deeper than 16 in., material costs increase significantly because web material other than #3 shorts must be bought (now you know where mill ends go).

After pricing a lot of wood-chord trusses here in the Houston area, I quickly found out that the size of a builder's truss order is the greatest single factor affecting the cost of trusses. If there are several truss fabricators in your area, each one may have a different business approach. One supplier will probably favor small-quantity, custom truss orders while another will specialize in mass production and large orders. Shopping around for the best price often means shopping for the fabricator whose business most fits yours.

Truss anatomy—A floor truss has only three components—chords, webs and connector plates. Each one is critical to the function of the truss. The wood chords, or outer members, are held rigidly apart by wood or metal webs. The strength-to-weight ratio of floor trusses is higher than that of solid-wood joists because the structural configuration of the truss converts the bending moments and shear forces (produced by loading) into compressive and tensile forces. These forces are directed through the individual truss members and transferred to walls.

Chords. The type of floor truss used most frequently in residential floor systems is called the parallel-chord 4x2 truss. As shown in the drawings on the facing page, the chords are evenly spaced from each other, and the designation "4x2" identifies them as 2x4 lumber with the wide surfaces facing each other. This configuration increases the structural efficiency of shallow trusses, and provides them with a larger bearing area on wall plates. It also gives trusses additional lateral rigidity to resist damage during transport and installation.

In residential construction, a floor truss will **most often bear on the underside of its bottom chord** (left drawing, previous page), just as a joist **bears on its bottom edge. But because of the** structural versatility of a truss, it can also be designed to bear on the underside of its top chord. In this case, the bottom chord is shortened and **the truss hangs between walls, instead of resting on top of them** (see right drawing, previous page). **This can be a useful feature when the overall** height of a building is restricted. But since height isn't usually a problem in residential construction, top-chord bearing trusses are seen more frequently in commercial work.

Chords can be made of spliced lumber as long as a metal connector plate is used to join the pieces. Chords are kiln-dried, and in the southern United States, they're generally made from No. 1 KD southern yellow pine. Truss fabricators commonly use either machine stress-rated lumber or visually graded lumber for truss chords, depending on the cost and availability of the lumber grade. During its early years, the wood-truss industry used only visually graded lumber, and trusses were designed conservatively to compensate for substantial variations in the strength and stiffness properties of the wood. Some time ago, engineers at Purdue University developed a technique to test and grade lumber by machine. Lumber graded by this process is called machine stress-rated (MSR) lumber, and truss fabricators are using it with increasing frequency for chord stock. Because every piece of MSR lumber is mechanically tested for stiffness and given a categorical strength rating, trusses can be designed to maximize the use of the wood's strength.

Webs. The members that connect the chords are called webs. Diagonal webs primarily resist the shear forces in the truss, and they are usually positioned at 45° to the chords. Vertical webs, which are placed perpendicular to the chords, are used at critical load-transfer points where additional strength is required. They also are used to reduce the loads going through the diagonal members.

Since the strength of the webs is not as critical as is the strength of the chords, a lower-quality wood can be used, such as #3 KD southern pine. Wood is the material used most frequently for webs, but metal webs are also used. Metal webs are stamped from 16-ga., 18-ga. or 20-ga. galvanized steel. Several truss manufacturers have designed the metal web to incorporate a connecting plate, thus reducing the number of pieces required to assemble the truss, as shown in the drawing below. Metal-web trusses provide greater clear spans for any given truss depth than wood-web trusses. The opening between the webs is larger, too, which allows more room for HVAC ducting. And they're lighter, so they are easier to install. But a major disadvantage of metal-web trusses is that a wide range of truss depths isn't always available. Also, metal webs are more susceptible to damage during transport and installation.

In a truss with wood webs, the ends of the diagonal webs are double-beveled wherever they meet a vertical web and the chord. There is an important structural reason for this cut. The objective in the design of any truss is for the centerlines of the chord and adjoining pairs of webs to intersect at one point. This is called the panel point. When centerlines don't intersect at

Types of metal webs

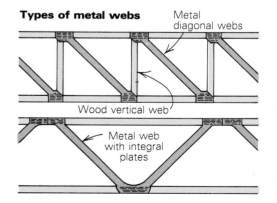

Metal diagonal webs

Wood vertical web

Metal web with integral plates

Floor trusses can be cantilevered over longer distances than wood joists. These trusses will support a balcony. Pressure-treated wood was used for the end web as a precaution against rot caused by possible water infiltration. The double vertical web member at the wall provides additional support for the deck loads on the cantilever.

a panel point, additional and undesirable stresses are introduced at the joint.

There are several different types of parallel-chord floor trusses, and what makes each one different is the arrangement of its webs. Depending on their arrangement, the webs will be either in compression or tension, and this dictates how loads are transferred from the truss to the structure supporting it. Pratt, Howe and Warren trusses were named after their respective developers. Howe and Pratt designed their trusses for railroad bridges. Two giant timber-framed trusses could span a river or a canyon, and trains would run over tracks that stretched alongside the bottom chords.

Connector plates. In 1952, A. Carol Sanford invented the toothed metal connector plate, which eliminated the need for nailing and gluing truss plates. Metal connector plates substantially reduced the cost of trusses by allowing them to be mass produced.

Variations of the toothed connector plate are now commonly used to assemble trusses. They are made from 16-ga., 18-ga. or 20-ga. hot-dipped galvanized steel, which is punched to form numerous metal prongs, or "teeth," that extend outward from one side of the plate. When embedded in the wood of the truss (usually by a hydraulic press), these teeth give the plate its holding power. Specifications for designing trusses using metal connector plates are available from the Truss Plate Institute, Inc. (583 D'Onofrio Drive, Suite 200, Madison, Wis. 53719), which also distributes test and research data.

If you've had a chance to examine connector plates, you may have noticed that they are not always the same size. As they increase in size, they provide more embedded teeth, and therefore more holding power. Larger plates are used in joints with higher stresses.

Deflection—The design of a truss is usually governed more by bending limitations than by anything else. Too much bending, or *deflection*, can make a floor noticeably springy and result in cracks in the finished ceiling below. The

American Institute of Timber Construction (333 W. Hampden Ave., Suite 712, Englewood, Colo. 80110) recommends deflection limitations for trusses of L/360 for the live-load portion of the total load and L/240 for the total load, where L is the span in inches. Live loads are calculated only if they are expected to be unusual—for example, placing a pair of pianos on the second floor. Generally, though, the total load calculation is used. Total load includes the weight of everything attached to or bearing on the top and bottom chords. It includes the dead weight of floor and ceiling materials, as well as the live loads of people, pianos and furniture.

To see how deflection limitations are used, suppose a floor truss must be selected to span 22 ft. between two walls. The maximum allowable total load deflection is calculated by converting the span distance to inches (22 x 12 = 264) and dividing this by 240. So a truss must be selected that will deflect no more than 1.1 inch at mid-span. In other words, if the truss is subjected to all the loads anticipated, the most it may deflect downward is 1.1 in. Remember that this is total deflection, including that caused by dead weight; it doesn't mean that every footstep will cause the truss to deflect 1.1 in.

Deflection limitations provide design parameters for the engineer in selecting an appropriate truss. Knowing the maximum deflection allowable over a particular span, the engineer can determine a truss depth that will deflect less than the maximum allowable distance. This can be done by mathematical calculation or by consulting data from truss-plate manufacturers concerning their products.

A floor system consists of the deck material and supporting network of trusses, and both components function together to transfer floor loads to the walls and foundation. But when trusses are engineered for a particular application, the effect of the deck material on truss stiffness is not taken into account. The floor-system strength is based solely on the stiffness of the truss itself. Because of this, the deflection calculations are on the conservative side.

Choosing a truss—Figuring the total loading on a floor system involves some calculation, but you don't necessarily have to hire an engineer or an architect to size trusses for your project. Engineering is provided either by the truss fabricator or by the plate manufacturer whose plates are purchased by the truss fabricator. These people work together to ensure that the combinations of plates, webs and chords are structurally sound.

To size a truss, the fabricator will first review your plans and look for any unusual circumstances that could affect the truss design. If nothing unusual is found, the truss design and depth will be chosen from reference manuals supplied by the plate manufacturer. These manuals list the standard truss designs and include span tables for each. The span tables simplify all the engineering variables into a very usable form, and are based on design formulas that come from actual laboratory tests of trusses. Spans are listed for four commonly used truss spacings: 12 in., 16 in., 19.2 in. and 24 in. The most common spacing for residential construction is 24 in., but trusses can be placed at any spacing that will allow the floor system to carry the loads specified in the building codes. All the spans listed in the tables assume the deflection limitations of L/360 or L/240.

Since truss-plate size is a critical factor in truss design, the plate manufacturer will often supply pre-certified engineering for the trusses, stating the limitations with in which they will meet certain performance standards, such as the deflection limitations.

If a standard design can be found to meet your needs, the truss fabricator will provide you with a certified drawing of it. Proof of this certification is the engineer's stamp on each page of the drawing.

If the design of your structure is unusual, with particularly long spans, heavy loads or unusual support conditions, it may call for a floor truss that has to be engineered specifically for your project. In this case you'll need to arrange for an outside engineer to do the calculations.

Typically, parallel-chord floor trusses with wood webs are available in depths ranging from 12 in. to 24 in., in 1-in. increments. The most common depths are 14 in. and 16 in. One plate manufacturer has designed a metal-web truss with the same actual depth dimensions as 2x8, 2x10 and 2x12 solid-wood joists. These smaller sizes are interchangeable with an ordinary joist-floor system.

The amount of space between webs is another thing to consider when you choose a truss. When ducts will be routed under the floor, the depth of the truss may be dictated by the size of the ducts. Usually this isn't a problem, since most fabricators have a standard truss design that includes a chase opening. To create a chase opening, one web in mid-span is removed to provide space for large ducts. The truss doesn't collapse on account of the missing web because shear forces are minimal at mid-span.

One last note about choosing trusses. Before you leave the fabricator's office, make sure that your order is complete. If you run short of joists, you can dash off to the lumberyard. But since trusses are precisely fabricated to your specifications, you will probably have to re-order the ones that you forgot. This can be expensive, and time-consuming as well.

Truss bearing details—A critical aspect of floor-truss design and installation is the amount and location of bearing surfaces, since the accumulated loads of the truss are concentrated here. Bearing details vary, depending on how and where the truss is being used.

For the simple-beam condition (drawing A, right), the truss is supported at each end, and rests on its lower chord, in joist fashion. It can also rest on its upper chord, and will carry the same amount of weight as an otherwise identical bottom-chord bearing truss.

A floor truss designed as a continuous beam will be supported on each end, as well as at one or more points in between. To ensure proper load transfer at the intermediate support points (a column or partition, perhaps), a vertical web member should be located directly over the bearing area (drawing B). Some truss fabricators will identify the intermediate load-bearing points on a continuous-span truss by stapling a green or red warning note to the area. These warnings are particularly helpful when you install the truss, since they help position the truss and keep you from accidentally installing it upside down.

When floor trusses are to be cantilevered (drawing C), to support a balcony, for example, it is particularly important that provisions be made for extra support at the bearing points (photo, previous page). The truss should always be designed so that a panel point rests over the bearing point. This transfers the load from the truss directly to the supporting element. To strengthen the area even more, the bottom and top chords can be doubled near the support point of the cantilever.

Trusses can also be used in multiples. In drawing D, two of them meet over an interior partition, while in drawing E, a steel beam provides the intermediate support.

Strengthening trusses—When floor systems are designed, the usual approach is to specify a single truss depth that will work for the entire system, regardless of the various span distances. This simplifies material orders and eliminates the possibility of mix-ups at the job site. Situations do arise, however, when it isn't practical or cost-effective to increase the depth of all the trusses in order to beef up just a few, and in these cases the truss fabricator can increase the stiffness of individual trusses.

Chords can be strengthened across the entire length of the truss by reinforcing the top and bottom chords with a second layer of wood, fastened into place with metal plates, nails or glue. Side-by-side floor trusses are often used to carry greater floor loads (bottom photo, next page), in a technique akin to sistering wood joists. Other options include using larger truss plates, stronger wood for the chords, or doubled webs in critical shear areas of the truss.

Sometimes floor trusses have a slight lengthwise curve that's built into them by the manufacturer. This curve is called *camber*, and it helps

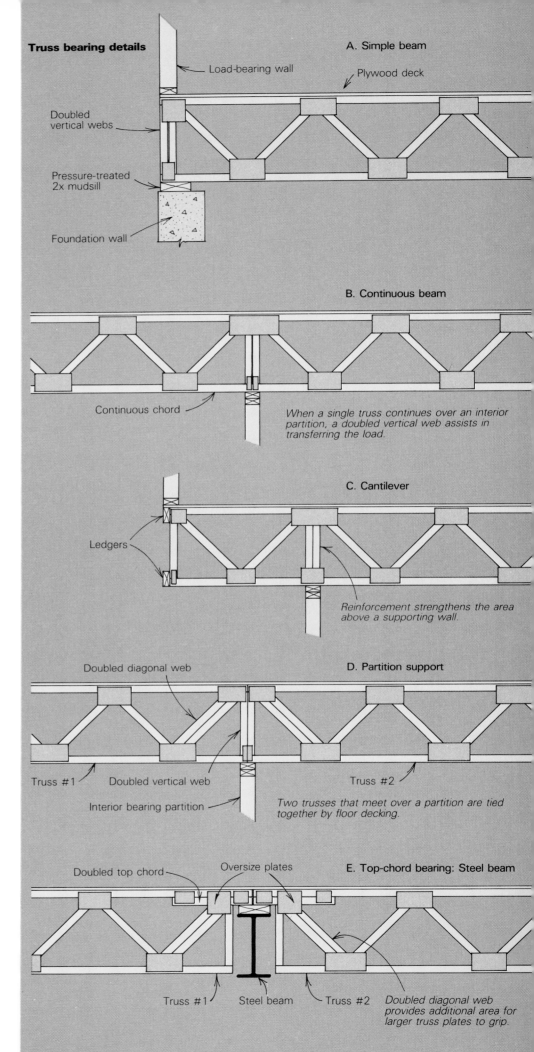

Truss bearing details

A. Simple beam

Load-bearing wall
Plywood deck
Doubled vertical webs
Pressure-treated 2x mudsill
Foundation wall

B. Continuous beam

Continuous chord

When a single truss continues over an interior partition, a doubled vertical web assists in transferring the load.

C. Cantilever

Ledgers

Reinforcement strengthens the area above a supporting wall.

D. Partition support

Doubled diagonal web
Truss #1
Doubled vertical web
Interior bearing partition
Truss #2

Two trusses that meet over a partition are tied together by floor decking.

E. Top-chord bearing: Steel beam

Doubled top chord
Oversize plates
Truss #1
Steel beam
Truss #2

Doubled diagonal web provides additional area for larger truss plates to grip.

Before the decking is nailed down, trusses should be measured at mid-span to make sure they haven't swayed out of alignment (note the tape measure in the photo). A pair of 2x4s can then be nailed to the trusses at the chase opening to ensure that the top and bottom chords stay at the proper spacing. These 2x4s are not strongbacks, however; strongbacks are usually 2x10s or 2x12s.

After the trusses are toenailed to the top plate, a ledger is nailed in place along their top corners. This further stabilizes the floor and provides an additional horizontal nailing surface for siding. To support additional loads, particularly at the edges of floor openings, trusses can be doubled.

too. They should be closely fitted. Truss-plate teeth should be evenly embedded into the wood for proper load transfer.

Installing floor trusses—A certain amount of care must be taken when installing floor trusses, but otherwise the installation isn't much different from the installation of wood joists. When trusses are delivered to the job site, be sure to check them against the design drawings. Any structural peculiarities, like bearing blocks or doubled chords, should be verified. A bearing block is the extra vertical web that's put in a truss over any support point. Watch for the identification tags on the truss to help you find the bearing blocks.

Metal-web trusses should be inspected carefully to ensure that no webs have been bent. As a general rule, most metal-web trusses are designed to put the webs in tension, so a bent web is not always a problem. But it's best not to take chances. Wood webs should be checked for looseness or damage, as should the plates.

As a rule, the truss fabricator will deliver the trusses to the job site. Make sure you have enough help on site to unload and place the trusses. Be careful when handling them because the edges of a sharp truss plate or metal web can cut. Wear gloves.

Some builders have told me that the reason they don't like to use trusses is because they have to hire a crane to lift large ones into position. I firmly believe that even large trusses can be manhandled up to first-floor wall plates, though the task is a little more difficult when you get to second-floor plates. If you decide to use a crane for the job, you might as well use it to stack the plywood up there, too. Just be sure to nail the trusses securely into place first, and be sure you don't overload them. Bottom-chord-bearing trusses can be further stabilized by temporarily nailing a stringer board to the ends of the trusses.

As the trusses are being set into place, make sure they're right side up. They're sometimes installed upside down by accident, and when this happens, webs and chords that were designed for compression will be in tension. Truss failure is the likely result of this mistake.

Once the trusses have been toenailed to the wall plate (if need be, you can nail through one of the holes in the connector plate) a ledger board (photo below left) is nailed to their top edge to stabilize them and provide a continous horizontal nailing surface for siding. The ledger fits into a notch created for this purpose by the truss fabricator.

If the top or bottom of the truss system is to be exposed when the structure is complete, it should be permanently tied together somewhere at mid-span. If the top chords are exposed to an unfloored attic, for example, the trusses should be braced with 2x material at least every 3 ft. along their length. If the bottom chords are exposed to an unfinished basement, bracing should not exceed 10-ft. intervals. □

to eliminate the visual effects of deflection and to control cracking. But camber does not add strength—it's simply a matter of appearance.

A common way to strengthen a floor system is to install a strongback, usually a 2x10 or 2x12, between the web openings to help distribute loading to adjacent trusses. Sometimes two 2x4s are slipped on edge through the web openings of a series of trusses at mid-span, and then nailed to a vertical web member, as shown in the top photo. This doesn't add much strength, but it helps to maintain truss alignment.

Choosing a truss fabricator—The strength of trusses is affected by the quality of their construction, so purchase your trusses from a reputable fabricator. If you aren't familiar with fabricators in your area, visit their production shops or look at samples of their work at job sites before you buy.

When you examine webs and chords, look at their alignment. Make sure that the centerlines of all the members intersect at a panel point, and that this point is adequately covered by a truss plate. Take a close look at the wood joints,

E. Kurt Albaugh, P. E., is a consulting engineer based in Houston, Tex. He has a patent pending for a new type of truss connector plate.

Laying out for Framing
A production method for translating the blueprints to the wall plates

by Jud Peake

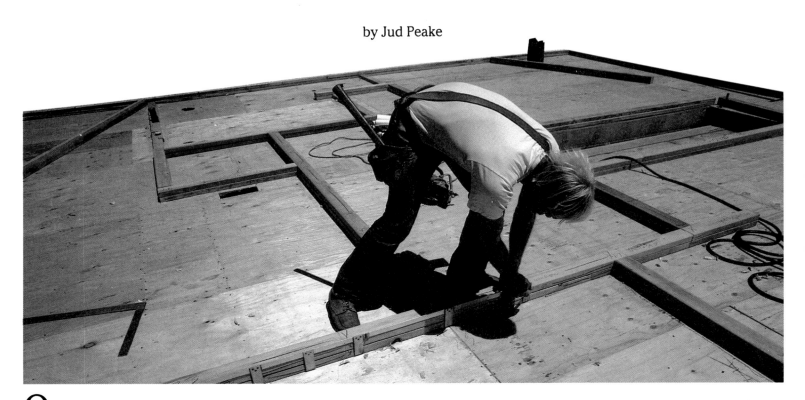

On big production projects where jobs are highly specialized, the carpenter assigned to layout uses a hammer very little. His tools are a lumber crayon (known in the trade as keel), a pencil, a layout stick, a channel marker and a tape measure. If he marks the top and bottom plates correctly, the carpenters may not even need a tape measure to frame up the walls. All of the figuring that involves the plans—and there are a lot of variables to anticipate—is done at the layout stage.

On large jobs, I don't necessarily frame the house I lay out. On houses I contract, I do both jobs. Either way, I treat layout as if there won't be anyone around to answer questions when the framing starts. Doing the layout as a separate operation increases the speed and accuracy of framing, whether you're building your own house by yourself or working as part of a big crew. The plans are an abstraction of the building to be constructed. Layout systematically translates your blueprints into a full-size set of templates—the top and bottom plate of each interior and exterior wall on a given level of the house. The pieces of the puzzle can then be cut and framed in sections.

There is little about a finished house that isn't determined by the layout. The actual procedure falls into four distinct steps. Within each step, you have to deal with wall heights

and the locations for windows, doors, corners, partitions, beams and point loads. You also have to deal with *specials* (which is anything else and usually means prefab components).

The first step in layout is to go over the blueprints and mark them up with the information you'll need, in the form it will be most useful. Step two is to measure out the slab or deck and establish chalklines representing every wall on that level. This is called *snapping out.* In step three you'll decide where walls begin and end and which ones will be framed first, and then cut top and bottom plates for each wall. This is called *plating.* Finally you *detail* the plates by marking them with all the information the framer will need to know to build the walls. Layout requires a thorough knowledge of framing. But even then, regional framing techniques vary widely.

Layout principles—Layout is based on parallel lines. If two lines are parallel and one is plumb, then the other will be plumb. Also, if a pair of lines meet at a right angle, then another pair of lines, each parallel to its counterpart in the first pair, will meet at a right angle. Stated less theoretically, put a 2x4 on top of another 2x4 and cut them to the same length. Using a square, draw a line across the edges of both every 16 in. Frame in between

these top and bottom plates with studs of equal length, and stand the wall up and plumb one end. Now all of the studs will be plumb. Doors and windows that have been laid out with these studs will be plumb. And if your foundation and deck are square and level, then any interior walls paralleled off the outside walls will also be square and plumb.

The second principle is to do all the layout at once. This way all the plates are fitted against each other, and you can be confident that the walls will work together before they grow vertically and become unmanageable.

The third principle is my own: avoid math. You can do most of your figuring in place. If you're looking for the plate and stud lengths of a rake wall (one whose top is built at the pitch of the roof it supports), snap it out full scale on the deck. Measure things as few times as possible. Once the walls are mapped out on the deck in chalk, don't measure them and then transfer this number to the plate stock sitting on sawhorses. Instead, lay the plate material right on the line and cut it in place. This saves time and reduces error—the kind that has to be fixed with a cat's paw.

Marking up the plans—Before you step out on the slab or deck, you need to go over the blueprints and systematically pick out the ele-

From *Fine Homebuilding* magazine (June 1984) 21:69-77

Marking up the plans

The first step in laying out is to add to the floor plan any further information needed to detail the plates later on. You need to write in the header length for each door and *tt* if it will require double trimmers. Windows show two numbers—the header length and the height of the sill jacks. You also need to mark beam pockets and the point loads under them, rake walls, and the locations of non-standard stud and header heights. The last notation is the direction the stud layout will run.

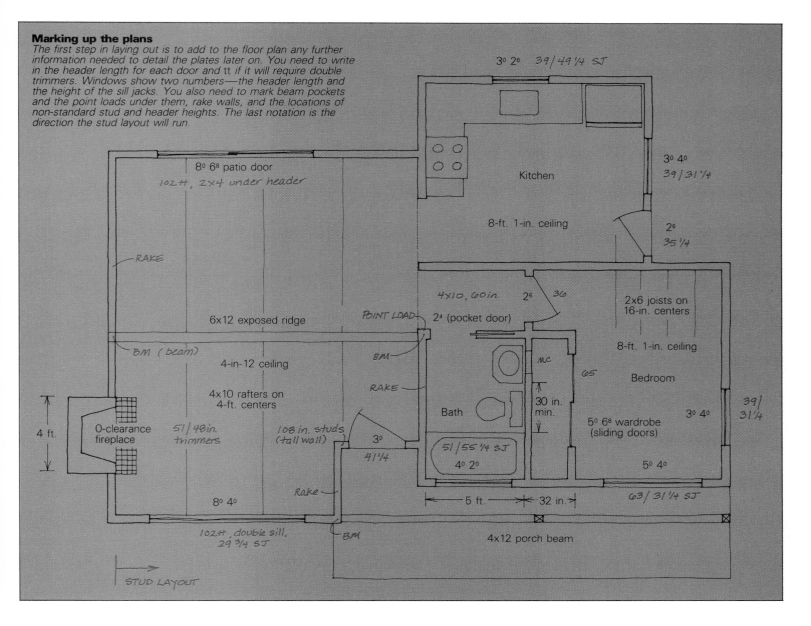

ments that should be part of the layout. I do this in the evening when I can slow down my pace a little bit. Thumb through the drawings just to review the general structure. Read the rough and finish carpentry specifications, and underline anything that isn't standard.

Next, go back to the first-floor plan in the blueprints. Write on the plans, next to the appropriate opening, the length of each door and window header. This will allow you to measure out and mark them on the plates later on without having to stop and figure, and will also allow whoever does the framing to make up a list and cut all of the header stock, sill and sill jacks at once. See sidebar on p. 34 for what these framing members are, and how to figure their lengths.

The drawing above shows a blueprint that contains many of the framing situations you will encounter. I'll use these same plans in explaining snapping out, plating and detailing. In this case, the blueprint is marked up with the necessary information for laying out. For doors, only the header length is written in; when nothing else is noted, this indicates a standard 6-ft. 8-in. door in an 8-ft. wall.

In the case of the aluminum patio door in

the living room, a 2x4 is needed to fur the header down to the right height. The *tt* next to the header length indicates double trimmers for this 8-ft. opening. The pocket door to the bathroom shows the length of the header, and its narrower width (4x10). Windows show two numbers: the first is the length of the header, and the second is the length of the sill jacks.

With the exception of the rake walls, which will be snapped out full scale on the deck, I have marked the only wall that doesn't use standard 8-ft. studs (the jog in the living room) with the notation *108-in. studs (tall wall)*. With a hodgepodge of elevations, it helps to mark up the plan view with colored pencil to differentiate stud heights.

The plans have also been marked for beam pockets (*BM*) on both ends of the ridge beam in the living room, and where the porch beam bears on the exterior wall. Resulting point loads (where significant loads have to be carried down to the ground) also have to be marked. The interior end of the ridge beam might not be picked up below without a notation. Also, studs should be doubled under joists that will be doubled on the floor above.

You'll also need to mark the dimensions of

any prefabricated items that will go into the house, like a medicine cabinet or roof trusses. (When using trusses, the building width can always err slightly on the narrow side, but don't ever make it too wide.)

One of the last things you'll want to note on the plans is at which end of the building you'll start the regular stud layout. The decision is yours. The side with the fewest jogs or offsets in the exterior walls is usually the place to start. When you reach a jog, compensate with the stud layout so that the studs, joists and rafters (or trusses) all line up, or *stack*. If the plans call for 2x6s on 2-ft. centers with trusses, code requires that they stack. It's a good idea anyway, both for increased structural strength, and for making multi-story mechanical runs like heat ducts easier.

In the drawing, I decided to pull my layout from the bottom left corner. By beginning there, the studs will pick up the rafter layout in the living room (exposed 4x10s on 4-ft. centers) and will work well with the ceiling joists and the porch overhang. You can also write down the cumulative dimensions of rooms, so that when you're snapping out, you'll be able to mark all of the intersecting

The Pieces to the Puzzle
An introduction to the components of a modern frame and how to size them

If you've done some framing, it takes only a second of looking at the detailing on a set of plates to know what the wall will look like when it's framed and raised. After that it takes only a pencil and a 2x4 scrap to make up a cut list of headers, sill jacks, rough sills, trimmers, channels and corners. But if you're new to how all of this goes together, you'll need to understand the different framing styles and the many exceptions created by using different kinds of windows, doors and finish.

Below I've explained how the basics work. Since I learned my framing in the West, my explanation will focus on how it's done there using basic production techniques, but I'll also point out more traditional methods as I go along.

Stud walls—Studs (2x4s or 2x6s on 16-in. or 24-in. centers) hold up the roof or floor above, and provide nailing surfaces at regular intervals for the interior and exterior finish. Precut studs are 92¼ in. long and are used to build a standardized 8-ft. wall that works economically with standard plywood and drywall sizes. The actual height of this wall is 8 ft. ¾ in. once you add 1½ in. each for the *bottom plate (sole plate)* that sits under the studs, and the two *top plates* (the *top plate* and the second top plate, called the *double top plate* or *doubler*) that complete the wall. Architects sometimes add to the confusion by specifying this same wall as 8 ft. 1 in. Wall studs are seldom shorter, but where economy isn't as important, they are often longer. If walls are over 10 ft., they will require *fire stops*—horizontal blocking that slows down the upward spread of fire.

To find stud height, check the elevations and sections in the blueprints. What will be listed here are wall heights, usually shown as finished floor to finished floor (*F.F. to F.F.*). In most cases, you can use the same dimension for rough floor to rough floor. To figure the stud length given this dimension, subtract 4½ in. (the thickness of the three plates), plus the subfloor thickness and joist depth.

Rake walls—These are also called gable-end walls, and they require the framing to fill in right up to the bottom of the pitched roof. This means that each stud will be a different length and will be cut at the roof pitch on top. Typically, a rafter will sit on the top of these walls. There are two ways to frame them. One is to build the lower part of the wall just as you do the walls under the eaves of the roof and then fill in the gables later. This is fine if there is a flat ceiling at the 8-ft. height. The other way is to build the wall in one unit with continuous studs. This is necessary for cathedral ceilings. See p. 36 for the specifics on figuring lengths and angles.

Wall intersections—When one wall intersects another, the framing has to provide a solid nailed connection between the two walls, and backing for the interior finish. The drawing above left shows how this is usually handled. Money can be saved by replacing backing studs with nail-on drywall corner clips, and by reducing corner units to a simple L, but there are disadvantages to these cutbacks.

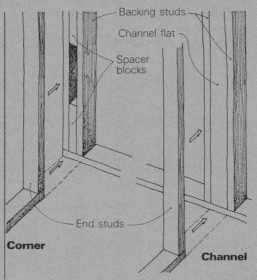

Corner — Backing studs, Channel flat, Spacer blocks, End studs

Channel

Doors—Door openings require *headers* to shift the weight of the roof in that area to both sides of the door. The vertical 2x support at each end of a header is called a *trimmer* (or *cripple*); the stud just outside of the trimmer that nails to the end of the header is called a *king stud*. The *rough opening* is the *rough-opening width*, measured between the trimmers, by the *rough-opening height*, which is measured from the floor to the bottom of the header.

Finding the height of the trimmers and the length of the header requires working backwards from the dimensions of the finished door. But you don't have to go through the full process each time—door headers should be 5 in. longer than the nominal width of the door. For example, a 37-in. header is needed for a 2-ft. 8-in. door. This 5-in. increment assumes that the trimmers will be framed very nearly plumb. Some carpenters use 5⅛ in. or even 5¼ in. to allow for sloppier framing.

Adding 5 in. to the header accommodates two 2x trimmers (1½ in. apiece, for a total of 3 in.) under the ends of the headers (drawing, above right). This leaves a rough opening 2 in. wider than the door. The remaining room is for two 1x side jambs (¾ in. apiece, for a total of 1½ in.), and ½ in. for shim space. (This leaves ¼ in. on each side.) Allow another ½ in. for exterior doors whose rabbeted jambs are closer to ⅞ in. thick. French doors will require closer to ½ in. of extra header length to account for the astragal (the vertical trim between the two doors that acts as a closure strip). If the door opening is wider than 8 ft., then the trimmers will need doubling, which requires another 3 in. of header length.

In the West, it's common to use 4x12 Douglas fir header stock in all 8-ft. walls. This system is fast, because all the framer has to do is cut the stock to length and nail it to the top plate—there are no *head jacks* to toenail, and you end up with a header at the right height that will span almost any opening. This system is admittedly wasteful, but gets around the cost of labor. Fir is still relatively inexpensive in large hunks, and the labor to cut and install the head jacks isn't.

If you aren't using 4x12s, check a span table for the correct header size (unless you are dealing with a non-bearing interior wall, where two flat 2x4s will do nicely). Typically, when

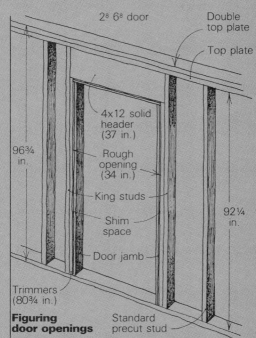

Figuring door openings — 2⁸ 6⁸ door, Double top plate, Top plate, 4x12 solid header (37 in.), Rough opening (34 in.), King studs, Shim space, Door jamb, 96¾ in., 92¼ in., Trimmers (80¾ in.), Standard precut stud

solid headers aren't used in 2x4 bearing walls, the choice is a laminated header made on site from two lengths of 2x with ⅜-in. plywood sandwiched between. Using either system, if a wall exceeds 96¾ in. in height, head jacks (cripples) will be needed between the header and the top plate.

Using a standard 6-ft. 8-in. door, the trimmers should be cut 80¾ in., no matter how tall the wall is, or what kind of header you use. Once the bottom plate is cut out within the doorway, this will leave a rough-opening height of 6 ft. 10¼ in. This height will accommodate the door (6 ft. 8 in.), the *head jamb* (¾ in.), and enough play for the finish floor and door swing. An aluminum patio door requires 1½ in. of furring under the header. Pocket doors and some bifold doors require an extra 2 in. of rough-opening height for their overhead tracks. In an 8-ft. wall, a 4x10 held tight to the top plate works nicely. Even if you are not using solid header stock, you'll need trimmers that are 82¾ in. for these doors.

Windows—A rough window frame is like a door opening with the bottom filled in—that's just how you frame one. The rough width of a window is measured between the trimmers. The rough height is measured from the bottom of the header to the top of the *rough sill*. This is a flat 2x, doubled if the window is 8 ft. or wider, that runs between the inside faces of the trimmers. Unless it's otherwise noted on the plans, windows are framed with the same height trimmers as the doors.

In some areas of the country, the trimmers are installed in two pieces (*split trimmer*, or *split jack*) with the rough sill cut 3 in. longer and sandwiched between. If double trimmers are used, then the inside pair can be framed this way. Underneath the rough sill, the stud layout is kept by *sill jacks* (or cripples), which are, in essence, short studs.

Finding the length of the rough sills and the sill jacks is fairly simple. Depending on how you deal with the trimmers, the length of the rough sill will be the same as the width of the

Figuring rough window openings

Wood windows used to be specified by lite sizes such as the double-hung 32/22 (two lites, each 32 in. wide by 22 in. high) at right. But these days, you are more likely to get a unit size like 37¾ in. by 53¼ in. (which includes sash and jamb allowances), or better yet, the rough opening of 38¼ in. by 53½ in. (which also includes the space for shimming). These numbers will allow you to figure the header, sill and sill-jack lengths so that the window will fit.

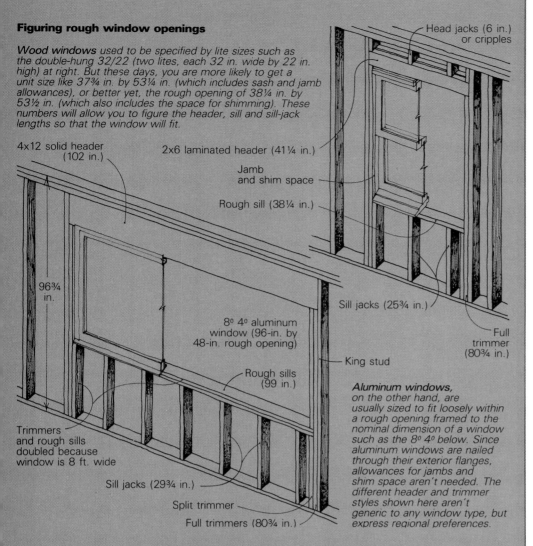

4x12 solid header (102 in.)

2x6 laminated header (41¼ in.)

Head jacks (6 in.) or cripples

Jamb and shim space

Rough sill (38¼ in.)

96¾ in.

8⁰ 4⁰ aluminum window (96-in. by 48-in. rough opening)

Sill jacks (25¾ in.)

Rough sills (99 in.)

King stud

Full trimmer (80¾ in.)

Trimmers and rough sills doubled because window is 8 ft. wide

Sill jacks (29¾ in.)

Split trimmer

Full trimmers (80¾ in.)

Aluminum windows, on the other hand, are usually sized to fit loosely within a rough opening framed to the nominal dimension of a window such as the 8⁰ 4⁰ below. Since aluminum windows are nailed through their exterior flanges, allowances for jambs and shim space aren't needed. The different header and trimmer styles shown here aren't generic to any window type, but express regional preferences.

---

rough opening or 3 in. longer. To get the length of the sill jacks, add the rough opening height to the thickness of the rough sill, and subtract this from the height of the trimmer.

The length of a window header has a lot to do with what kind of window will fill the hole—aluminum or wood. To get the header length on an aluminum window less than 8 ft. wide, just add 3 in. to the nominal window width to allow for a trimmer on each side (drawing, above). This is because aluminum windows are generally manufactured to fit loosely in a rough opening of their nominal size. Check with your supplier to be safe. The first number stated for windows (and for doors) is the width— a 3⁰ x 5⁰ aluminum window needs a rough opening 36 in. wide and 60 in. high. For windows wider than 8 ft., you'll need to use double trimmers, which will increase the length of your header 3 in.

Wood windows were once specified by the size of the lite, or glass, which didn't account for the sash or frame that surrounded it. The rules of thumb about how many inches to add for the sash to get the rough-opening size are quite general because the width of stiles and rails differs among manufacturers. The rules are also specific to window type (double-hung, single or double casement, sliders, awning and hopper) since the number and width of their stiles and rails vary.

Today, most manufacturers' literature gives a unit dimension (this includes sash and frame) and a rough opening dimension. Use the rough-opening measurements for laying out, and add 3 in. to get the correct header

length on windows under 8 ft.; 6 in. over 8 ft. If the rough-opening size isn't specified in the plans or by the manufacturer, you've got to measure the windows on site. If they are set in jambs, add 3½ in. (two trimmers plus ½ in. for play) to get the correct header length; if not, add 5 in., as you would for a door.

Specials—This catchall category includes any member or fixture in the frame that's big enough to worry about. Medicine cabinets and intersecting beams are *specials*. Toilet-paper holders aren't; they can be dealt with later.

Your main concern with most specials is to provide backing for the interior wall finish, and a break in the regular stud layout if an opening is required. Beam pockets are simple. Imagine a ridge beam that is supported on one end by a rake wall. There are a number of ways to frame the bearing pocket for the beam, but any of these schemes should include a post or double stud beneath the beam, and backing on either side of it for nailing wall finish or trim.

Tubs, showers and prefab medicine cabinets require both backing and blocking. Blocking is what it sounds like—short horizontal blocks or a 2x let into the studs for nailing the wall finish. Backing is usually an extra stud and does the same thing only vertically. In the case of a corner bathtub, then, the line of blocks that sits just above the lip of the tub where the drywall will be nailed is the blocking. At the end and side of the tub where the wallboard will butt, you'll find an extra stud for backing. —*Paul Spring*

Snapping out. **Over (p. 36): Peake uses his foot to anchor one end of the chalkline while snapping out a short interior partition. The red keel X by his right foot tells the framer on which side of the line to nail down the wall once it's raised. This project was unusual because the exterior walls were framed, sheathed and raised before the interior was laid out.**

interior walls along one side of the building by pulling the tape from just one point rather than having to do this room by room.

Snapping out—Snapping out isn't much more difficult than redrawing the architect's floor plan full size on the deck. Using a chalkbox, you need to snap only one side of each wall and draw a big X with keel every few feet on the side of this line that the wall will sit.

Measure for the outside walls first. Come in 3½ in. (for a 2x4 wall) or 5½ in. (for a 2x6 wall) from the edge of the deck. Be sure to measure from the building line, not from the edge of the plywood, in case it's been cut short or long. Don't even trust the rim joists without checking them with a level for plumb, since they may be rolled in or out.

Snapping lines should go quickly. To hold the end of your string on a wood deck, just hook it over the edge of the plywood, or use a nail or scratch awl driven into the deck. If the slab is very green (poured less than a week before), a drywall nail will usually penetrate the concrete; if not, use a concrete nail. You can even hold the string with one foot if the wall that you're snapping is short (see photo on p. 36).

On very long walls, especially on windy days, have someone put a finger or foot near the middle of the line and snap each side separately. This will keep the chalkline truer. If you're working alone, close the return crank on the chalkbox to lock it and hook it over the edge of the deck so it's secure on both ends.

I use red chalk for layout. Lampblack also shows up well, but blue isn't a good choice in my area because it's the favorite of plywood crews. When I expect rain or even heavy dew, I use concrete pigment (Dowmans Cement and Mortar Colors, Box 2857, Long Beach, Calif. 90801) instead of chalk. If you're doing a lot of layout, get yourself a couple of chalkboxes with gear-driven rewind; if you've made a lot of mistakes, correct them with a different color chalk. When it comes to keel, I use red mostly; blue has a way of disappearing.

In snapping out the exterior walls, don't be concerned about intersecting lines where the walls come together. This problem will be solved when you do the plating. Just concentrate on getting the lines down accurately. Once you've done that, check the lines for square by measuring diagonals, and check the dimensions again carefully. On slabs and first floors, check to see that where you have put the exterior walls will allow the siding to lap over the concrete for weathertightness.

I usually snap out rake walls on the deck (see sidebar, p. 36). This will let the framer cut the studs in place between the chalklines, once again avoiding math. Make sure that you

Rake walls

Rake walls are fairly simple to plate and detail if you snap them out full scale on the deck, and keep in mind how they will relate to the rafters that will eventually sit on them. The bottom plate of the wall will look like any other. The trick is locating the top plate so the framer can fill in the rake-wall studs without doing any calculations.

In this case the rake wall will intersect a standard 96¾-in. wall. The first real complication is dealing with the bird's mouth on the rafter. With 2x4 walls, I use a 3½-in. level cut so that the rake wall dies into the 8-ft. wall at the top inside edge of its double plate. This also allows me to measure the run of the rake wall from the inside of the 8-ft. wall to the near face of the ridge beam. In this case, that's 10 ft. Since the pitch is 4-in-12, the rise between those points is 40 in.

Now back to the deck. Measure along the 8-ft. wall, 96¾ in. from the chalked *baseline* of the rake wall. This baseline was snapped out like the other wall lines as a guide for positioning the inside edge of the bottom plate once the wall is framed and raised. But it is also a convenient starting point for the full-scale elevation of the rake wall that you are going to snap out on the deck. In this case, it will be used to represent the bottom of the bottom plate.

The next step is to lay out the

bird's mouth of the rafter above the 96¾-in. mark you just made along the 8-ft. wall. Then take two pieces of 2x scrap and lay them inside the line at approximately a 4-in-12 pitch to represent the rake top plates. The reason you use two pieces of 2x scrap instead of just subtracting 3 in. for the rake top plates is that the vertical thickness of these plates when they are at a pitch will be greater than when they are horizontal. In the case of a 4-in-12, two plates add up to about 3¼ in.

Now make a mark on the chalkline of the 8-ft. wall just below the two scraps. This point represents the top (short point) of the shortest stud in the rake wall, and will be used to establish the line of the rake top plate. To complete this line across the deck, you need to create a large 4-in-12 triangle. In this case, a 4-ft. leg and a 12-ft. leg work out nicely. Actually, this triangle can be any size as long as its proportions are correct, it is close to the length of the rake wall, and it is positioned so that its hypotenuse represents the bottom of the rake top plate.

To lay out this triangle, first make a mark on the rake baseline 12 ft. out from the inside of the 8-ft. wall. From this point, measure up 93½ in. (the height of the 8-ft. wall less the thickness of the two top plates) parallel to the 8-ft. wall, plus another 4 ft. to take care of the rise. Now use a snapline to connect

this point with the one on the 8-ft. wall that you established below the scrap top plates. This is the hypotenuse of the 4-in-12 triangle, and as the bottom edge of the rake top plate it will be used by the framers to cut the tops of the rake studs to length.

To establish the top end of the rake wall, find the point along the rake baseline that represents the near face of the ridge beam. When the rake wall is framed, the top plates will die at the top inside face of this beam. Now snap a line that is parallel to the 8-ft. wall that starts at this point on the rake baseline and ends by intersecting the rake top plate line you just snapped in the center of the deck.

To check your layout, you can use your tape measure to get the height of the longest stud in the rake wall at its long point. Do this by measuring along the chalkline you just established at the inside face of the ridge beam. Stretch your tape from the baseline of the rake wall to the bottom of the top plates, and then subtract 1½ in. for the bottom plate. This measurement should be the same as the shortest stud plus the rise of 40 in. that was figured earlier.

To finish up, determine the position of the ridge beam by laying out the bird's mouth at the top. By subtracting the depth of the ridge beam, you can also determine the length of the posts beneath it.
—*J. P.*

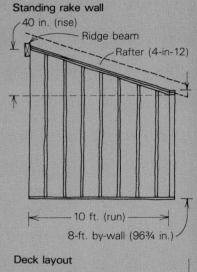

Standing rake wall

40 in. (rise)
Ridge beam
Rafter (4-in-12)
10 ft. (run)
8-ft. by-wall (96¾ in.)

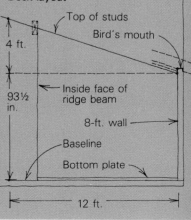

Deck layout

Top of studs
Bird's mouth
4 ft.
Inside face of ridge beam
93½ in.
8-ft. wall
Baseline
Bottom plate
12 ft.

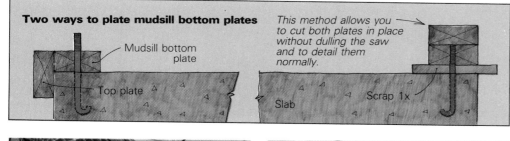

Plating the deck

All of the plates here above have been cut and tacked flat on the deck. This way they check themselves by butting tightly at every intersection. To get to this point, you have to decide which walls run by, and which ones butt. By-walls are framed first, and the longest walls are typically chosen for by-walls. Usually, parallel walls are then designated as by-walls. Rake walls should be framed as butt-walls, and this leads unavoidably to one living-room wall being plated log-cabin—one end running by and the other end butting.

This wall necessarily plated log-cabin to keep the intersecting rake wall a butt wall, and to keep from breaking up the continuous kitchen end-wall into two short sections.

BOTTOM OF PLATE (baseline)   TOP OF STUD

Rake plates should be plated as butt-walls.

This wall is plated by because it runs through to the exterior and is parallel with other by-walls.

These walls are plated as butt-walls because they run perpendicular to long by-walls.

Long parallel walls should be designated as by-walls.

These plates are up on edge and out of position because of stubbed-up plumbing from below.

Two ways to plate mudsill bottom plates

Mudsill bottom plate

Top plate

This method allows you to cut both plates in place without dulling the saw and to detail them normally.

Slab   Scrap 1x

Plating. Peake cuts in the top and bottom plate of a small raceway (left), and tacks each of the two sets of plates together with 8d nails (right) so they can be detailed. One of the first decisions in plating is which walls will run by, and which will butt. Here the long interior wall is a by-wall, and the parallel wall of the raceway is plated the same way.

let the framers know with directions in keel whether your lines represent the top and bottom of studs, or the plates themselves.

Once the exterior walls have been chalked out, snap the interior walls by paralleling them (measuring out the same distance from several points and connecting them). When a wall ends without butting another wall, indicate where with a symbol that looks like a dollar sign with a single vertical bar. Follow the written dimensions given on the blueprints. Before you use an architect's scale to get a missing dimension, make sure that you can't find it by adding and subtracting others. If a mistake is made—whether it's yours or the draftsman's—you're the one who's going to fix it. Yet it's always better if what you've done reflects the approved plans.

Occasionally, though, you will have to adjust room sizes to accommodate some unanticipated condition. The rule of thumb here is to adjust in the largest rooms. The sizes of smaller spaces are usually dictated by the building code or some prefabricated item.

Plating—Plating involves cutting a top and bottom plate for each wall, tacking them together and laying them in place on the deck. To do this, you must decide at each intersection of walls which ones run through (*by-walls*) and which ones stop short (*butt-walls*). Walls that should be framed first—usually the longest exterior walls—are designated by-walls. Usually, the walls parallel to these will also be plated as by-walls. The framing and plating is simpler if you plate rake walls as butt-walls since the top plates are detailed in place (on their full-scale layout lines) near the middle of the deck. If plated as a by-wall, the top rake plate would lap over the plates of exterior walls that run perpendicular.

What you want to try to avoid in plating is *log-cabining*—building walls that run by at one corner and butt at the other. Such walls will probably have to be slid into position after they're framed and raised, which isn't easy, especially with a heavy wall.

The drawing top left shows the plating for this floor plan, and lists reasons why some walls have been designated butt-walls and others by-walls. Notice that the living-room wall at the top of the drawing has been plated log-cabin. So much for hard-and-fast rules. In this case, the rake wall at the end of the living room had to be a butt wall, and the kitchen wall on the other end had to run by so it wouldn't end up in two extremely short sections. The result is necessary log-cabining.

To do the plating, spread the top and bottom plate stock near the snapped lines. Use long, straight pieces. A crooked top plate can drive a framer crazy when it's time to straighten the raised wall with braces. To make the length on long walls, top plates will have to butt together at the center of a stud. The middle of a solid header is an even better spot. Breaks in the top plate and the double plate (the framer will be supplying this permanent tie between walls) have to be at least 4 ft. apart. That means to stay at least 4 ft. away

from intersecting walls when laying out a break in the top plate, since the double plate of this wall will have to end there.

Very long walls will have to be framed in sections. With an average number of headers in an 8-ft. high 2x4 wall, each carpenter on the site should be able to handle at least 10 lineal feet of wall when it comes time to raise it. If you're going to break a long wall into separate sections, you'll need to end the top plate at the center of a stud. The bottom plate should be broken at the same place.

The plate stock for the bottom plate should be tacked flat to a wood deck along the snapped line with an 8d nail near each end. It's fine if it laps over nearby wall lines, because the next step is to crosscut it in place by eyeballing the chalkline of the intersecting wall. Now lay down the top plate in the same way, cut it and tack it to the bottom plate with two more 8d nails (photos previous page). The only exception to nailing the plates together is a rake wall where the top plate will be left on the deck on its angled layout line.

In the case of a slab, the bottom plate will be a mudsill. I use a bolt marker like Don Dunkley's (text and drawing, p. 21). I usually make mine out of a piece of 1x2 with a joist-hanger nail at 3½ in. and at 5½ in. Once the mudsill is drilled out and set on the bolts, you have to deal with the top plate, which won't tack down to the mudsill because of the bolts.

Some carpenters hang the top plate off the edge of the mudsill (drawing previous page, center left), and then detail the layout across the edge of the top plate and the flat of the mudsill. But the framers will like it better if you shim up the mudsill with 1x scraps (drawing previous page, center right). This allows you to cut both the mudsill and the top plate in place without dulling your sawblade on the concrete, and to detail them normally.

Detailing—After all the plating is complete, detailing can begin. It is done in three stages: recording the information that you've added to the blueprints on the plates, marking out the precise measurements for headers, corners and intersecting walls on the face and edges of the plates, and then adding the stud layout. The drawing on the facing page shows the floor plan with the plates fully detailed.

Layout style varies widely from region to region. One difference is in detailing shorthand. Layouts often contain more detailing than is really necessary. For instance, if you indicate where the end of a header falls with a line, and then make an *X* for the king stud beyond the line, it will be evident to the framer that the trimmer goes on the other side of the line. Writing *T* for trimmer (or *C* for cripple, de-

pending on the terminology you use) takes time, and doesn't add any information.

Another difference is the orientation of the plates. Some production carpenters tack the plates together and then toenail them along the chalkline with their edges up, rather than flat. But there are several advantages to running them flat. The first is that they check themselves. They can't be too long or too short because they are laid in the precise positions that they will occupy once they have been framed. Second, the location of headers and wall intersections is easy to see when it's detailed on the top of a flat plate, and won't get overlooked or misframed. Last, all the information necessary for the framers to cut the double plates that interconnect the walls is

marked on the surface to which they will be nailed—the top face of the top plate.

Almost all the marks that you'll put on the plates will be on the top of the top plate, and on one set of edges. The way to determine which set of edges is to approach a pair of tacked-down plates as if you are going to frame the wall. The trick is to figure out from which direction the top plate will be separated, the studs added and the wall raised. With an exterior wall this is easy. The top plate will be walked to the interior of the deck with its top markings facing the opposite side and its stud layout (which is on the edge) up. This way, once the vertical members have been filled in, the wall can merely be tilted up into place without having to reorient it. Exteri-

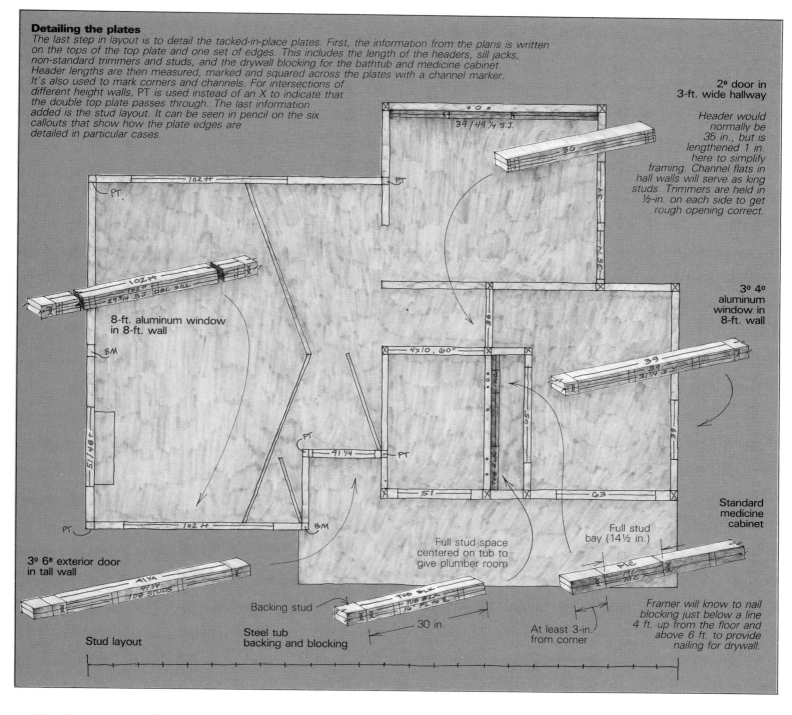

Detailing the plates

The last step in layout is to detail the tacked-in-place plates. First, the information from the plans is written on the tops of the top plate and one set of edges. This includes the length of the headers, sill jacks, non-standard trimmers and studs, and the drywall blocking for the bathtub and medicine cabinet. Header lengths are then measured, marked and squared across the plates with a channel marker. It's also used to mark corners and channels. For intersections of different height walls, PT is used instead of an X to indicate that the double top plate passes through. The last information added is the stud layout. It can be seen in pencil on the six callouts that show how the plate edges are detailed in particular cases.

2⁶ door in 3-ft. wide hallway

Header would normally be 35 in., but is lengthened 1 in. here to simplify framing. Channel flats in hall walls will serve as king studs. Trimmers are held in ½-in. on each side to get rough opening correct.

8-ft. aluminum window in 8-ft. wall

3⁰ 4⁰ aluminum window in 8-ft. wall

Standard medicine cabinet

Full stud space centered on tub to give plumber room

Full stud bay (14½ in.)

Framer will know to nail blocking just below a line 4 ft. up from the floor and above 6 ft. to provide nailing for drywall.

3⁰ 6⁸ exterior door in tall wall

Stud layout

Backing stud

Steel tub backing and blocking

30 in.

At least 3-in. from corner

---

or plates, then, are typically detailed on their outside edges. Interior partitions, since they can often be tilted up from either direction, require the layout carpenter to guess how and where the framers will set up.

You're now ready to do the detailing. Forget your tape measure for the moment, take the plans in hand and walk around the deck copying the information you've written there onto the plates. For each window and door, write its header length on the top of the top plate near where the header will be nailed, and *tt* if it requires double trimmers. Now lean over the plates and mark the outside edge of the top plate with the same information.

As long as you detail the rough-opening dimensions of the windows on the plates when you lay out, you don't have to figure the height of the rough sill or the length of the sill jacks. The framer can do this by measuring

the rough-opening height down from the header, installing the rough sill, and then filling in with the sill jacks. However, if you figure the length of the sill jacks at the layout stage, they can be precut for the framing. Write this length in keel on the outside edge of the bottom plate and follow it with the letters *SJ*. For any walls with studs longer than the standard 92¼ in., write their lengths on the face of the top plate and on the outside edge of one of the plates. Note the location of prefab items, like medicine cabinets *(MC)*, and drywall blocking around fixtures like bathtubs *(TUB BLK)*. Label rake walls *RAKE*.

At this point you can begin to mark corners and channels on the top and edges of the plates. Both of these are just wall intersections—one at the end of a wall, one in the middle—and are drawn the same width as the plate stock. When these intersections are

framed, they'll need backing studs for wall finish, but since corners and channels are usually made up as units that include their backing studs, you need only to show the intersection of the plates for the framer to get it right.

To detail the corners, use a pencil to scribe a line onto the inside edge of the by-wall plates along the side of the butt-wall plates. Then use a channel marker (see the box on the next page) to continue that line across the face of the top plate and down the other edges. These lines will show exactly where to nail the walls together once they are framed and raised. Channels aren't much different from corners, except you have two lines to scribe—one on each side of the intersecting plates. Put a big keel *X* between the lines on the face of the top plate, and on each of the outside edges of the plates to show where the double plates of the intersecting wall will create a half-lap. If the walls connect at different heights, the double plates of the by-wall shouldn't be broken for the double plate of the butt wall. The framer should be warned by marking the corner with the letters *PT*, which tell him to *plate through*.

A common framing mistake, usually discovered after the wall is raised, is putting the flat stud of the channel on the wrong side of the wall from the intersecting partition. As long as the framer knows the flat-plating method used here and doesn't reverse the top and bottom plates, where to locate the flat will be obvious. The key is the *X* that marks the channel. Because you are prevented from making any marks on the inside edges of the by-wall plate when you scribe the butt-wall plate to it by the plates themselves, the *X* will get marked only on the opposite set of edges from where the intersection will actually happen. So the framer should nail the channel flush with the edge of the plate that doesn't have an *X*.

Now detail the window and door openings. Following the blueprints, measure accurately to each end of the headers and use your channel marker to square the lines across the top plate and down the outside edges. Make an *X* on the outside of each of these lines to indicate the king stud.

When making an *X* over the edges of the plates, you can save yourself an extra motion by making two intersecting half-circles. This will leave an *X* on each plate when they are separated. The only time I show the location of trimmers is when they are doubled, which I indicate with *tt* (see the 8° 4° living-room window callout in the drawing, previous page).

Interior doors are often placed near the corner of the room they serve. The standard way to frame them is to let the king stud act as one of the backing studs in the channel, compressing space. Once the drywall is hung, this leaves a little less than 3 in. for casing. If the casing is wider than this, it will have to be scribed to the wall. If the space is even narrower and the door is in a butt-wall such as the 2-ft. 6-in. door in the 3-ft. hallway in the plan, you'll have to use the channel flat as a king stud. A useful rule of thumb is that the space left for the trim is about the same as the distance from the studs of the intersecting wall to the inside face of the trimmer.

There are a few special items in bathrooms that need detailing. The medicine cabinet fits between studs, but it will need blocking above at 6 ft. off the floor and below at 4 ft. If the medicine cabinet is near a corner, double the end stud that nails to the channel to give the necessary room for the swing of the cabinet door. Bathtubs and showers should be blocked along their top edges. Detail this by specifying a height on the plate that runs from the floor to the centerline of the blocking. A double stud or flat stud should be laid out to pick up the side and end of the tub or shower. In the drawing, you can see that I've also centered a standard stud space on the plumbing end of the tub to make it easier for the plumber to run supply lines and a drain.

Beam pockets can be detailed with your channel marker, but label them *BM* so they are not confused with channels. The posts under these beams are detailed with their actual width marked on the edge of the plates with keel, and their nominal size written in between these lines. Also give a length for the post if it is different from stud height.

Stud layout—The regular stud layout comes last and is done on the outside edges of the plates. Pull all the outside and inside walls that run perpendicular to the joists and rafters from the same end of the building. Do not break your layout and start again at partitions, but continue the full length of the wall. The standard 16-in. and 24-in. centers are meant to work modularly with 4-ft., 8-ft. and 12-ft. sheet materials. Remember that 16 in. o. c. means from the end of the building to the center of the first stud, so reduce your layout by ¾ in. each time when pulling from the corner (15¼ in. to the first stud, 31¼ in. to the second stud, and so on). This way the plywood sheathing and subfloor will work out with a minimum of cutting and waste.

It's a common mistake to have drywall on your mind when laying out studs. Drywall is relatively cheap, easy to cut, and can be bought in 12-ft. lengths. It should be at the bottom of your list of worries when laying out.

Stud layout should be done in pencil. If you are using a layout stick, you can put your tape measure back in your nail bag once you get started. Scribe along both sides of each finger of the layout stick to mark for the studs, then reposition it farther down the plate and repeat. If you aren't using a layout stick, stretch your tape the length of the wall and make marks at 15¼, 31¼, etc. Then come back with a combination square set at a depth of 3 in., square these marks down the outside edges of the plates, and make an *X* on the leading side of the line. Don't bother to draw a line for both sides of the stud, but don't lose your concentration either when making the Xs. Putting them on the wrong side of the line will cause big headaches later.

Rake top plates can be laid out by stretching a tape from the by-walls they butt, but you'll have to hold the tape perpendicular to the by-wall and keep moving the end of it farther down the by-wall so that the stud centers on the tape will intersect the angled rake plate. A better method is to measure the distance between studs along the angled plate after marking the first few, and then use this increment to mark the studs thereafter.

The last thing to do before you leave the deck is to look over the tops of the plates and make sure that every room has a door in it (a common but embarrassing mistake), and that all the channels are marked. These marks are easy to spot because they are on the top of the plates. Finally, to get the framing off to a good start you can cut all the headers, sills, sill jacks, and specials, as you already have a cut list on the marked-up plans. □

Jud Peake is a contractor and a member of Carpenters Local 36 in Oakland, Calif.

Tools of the trade

Most production layout tools were born of necessity on the site and made with available materials on a rainy day. One such tool is the channel marker (middle photo), a simple square made out of short pieces of plate stock and used for outlining corners and channels. It should have a leg 3 in. long (the depth of two 2x plates) and another leg 3½ in. long (the width of a 2x4). Both legs are 3½ in. wide. I make a more durable version with aluminum flat stock that includes a 1½-in. flange at the top. By turning the square over, you can lay out the thickness of a stud with this flange.

Two more tools that will speed things up are a layout stick (top photo) and a keel/pencil holder (bottom photo). This last item is just a short piece of ½-in. clear plastic tubing that will take a carpenter's pencil in one end, and your keel in the other. Layout sticks can be made out of standard aluminum extrusions riveted together. The 1½-in. wide and 3-in. long fingers on mine are laid out for 16-in. centers and 24-in. centers. I even threw a hinge into my stick so that it could fold up to fit in a standard carpenter's toolbox. —*J. P.*

Stud-Wall Framing

With most of the thinking already done, nailing together a tight frame requires equal parts of accuracy and speed

by Paul Spring

I spent much of my former career as a carpenter building a reputation for demanding finishwork, but some of my best memories center around the sweaty satisfaction of slugging 16d sinkers into 2x4 plates as fast as I could feed the nails to my hammer.

The emphasis in framing is on speed. A lot has to happen in a short time. Accuracy, however, is no less important. The problems created by sloppy framing—studs that bow in and out, walls that won't plumb up and rooms that are out of square—have to be dealt with each time a new layer of material is added.

The fastest framing is done using a production system. But these techniques have long been the domain of the tract carpenter, and bring to mind legendary speed coupled with a legendary disregard for quality. However, production methods don't have to dictate a certain level of care. Instead, they teach how to break down a process into its basic components and how to economize on motion.

Done well, production framing is a collection of planned movements that concentrates on rhythmic physical output. It requires little problem-solving since most of the head-scratching has been done at the layout stage. As long as the layout has been done with care, a good framer can nail together and raise the walls of a small home in a few days, and still produce a house in which it's a pleasure to hang doors and scribe-fit cabinets. And this pace will give both the novice and the professional builder more time on the finish end of things to add the finely crafted touches that are rare in these days of rising costs.

If you know what the basic components of a frame are, nailing the walls together is simple. If not, you'll need to read the article on layout on pp. 32-40. After figuring out which walls get built first, you will separate the bottom and top plates (which were temporarily nailed together so that identical layout marks could be made on them); fill in between with studs, corners, channels, headers, sills, jacks and trimmers; and then nail them all together while everything is still flat on the deck or slab. The next step is to add the double top plate and let-in bracing. Finally you'll be able to raise the wall and either brace it temporarily or nail it to neighboring wall sections at corners and channels. Before joists or rafters are added, everything has to be *plumbed* and *lined*—this means racking and straightening the walls so that they are plumb and their top

Walls can be framed with surprising speed if they're laid out well. The pace isn't frantic— it's rhythmic, and based on coordination, economy of motion and anticipation.

plates exactly mimic the layout that was snapped on the deck—as discussed in the article on pp. 50-53.

For the sake of simplicity, I have stuck largely to giving directions for nailing together a single exterior 2x4 wall with most of the usual components. I have tried to mention how this process would be different under different circumstances, and how each section of the wall is part of a larger whole. If you are using 2x6s or have adopted less costly framing techniques, such as the ones suggested by NAHB's OVE (Optimum Value Engineering), you'll have to extrapolate at times from the more traditional methods explained here.

Getting things ready—The carpenter I apprenticed to would begin wall framing on a Thursday so he could recover over the weekend from those first two grueling days of keeping up with the hot young framers. You don't need to plan things to this degree, but

you do need to make sure that the right tools and materials are at hand. Leave the detailed plates tacked in place on the deck for the moment (or up on foundation bolts in the case of a slab). Make sure the deck is clean; if it's not, sweep it. This surface is going to be the center of your universe for the next few days, so keep it spotless and plan how you will use each inch of it.

While you are setting up on the deck or slab, a helper can be cutting headers to length if this hasn't already been done. If the ground is flat, set up on sawhorses; if not, use one corner of the deck or slab. A cutting list can be made directly from the layout on the plates. I usually number each door or window opening sequentially around the deck with keel (lumber crayon) on the top plates, and then use these numbers to identify the headers as I cut them. This way I can easily find the piece I need for a particular wall and snake it out of the pile of corners, trimmers, channels and headers stacked on the deck. The information for cutting rough sills, sill jacks, and blocking is right there on the layout too, and your helper can make up a package for each opening.

If you're working by yourself or with only one other carpenter, it's just as easy to cut this 2x material in place when you're framing. The only exception is trimmers, which can be counted up and gang-cut if the headers are all to be at the same height. If you aren't using standard 92¼-in. precut studs, now is also the time to cut studs to length. Gang-cutting them is easiest, but precision at this stage is still important. If a few studs cut short of the line happen to get nailed next to each other in a wall, a dip will be left in the floor above.

You should also count up the total number of corners and channels you will need, and nail these together up on the deck or slab. Many framers don't bother with this step. They nail their corners and channels together when they're framing, but pre-assembly avoids having to sneak edge nails into a channel that faces down and is crowded by a regular layout stud.

Before you litter the deck or slab with any more material, you first must figure out how much of the frame you are going to nail together before you raise some walls, and where you are going to begin framing. On some second-floor or steep-site jobs where the plan is very cluttered, it pays to stack the frame. This

means nailing together all of the walls and then raising them at one time. But this requires a lot of planning since the walls literally have to be built on top of each other (often three walls deep), and a big enough crew to lift and carry the walls into place. Big production jobs use framing tables—essentially huge wall jigs—or the flat ground around the house to complete all of the walls on one level before raising them. But usually it's best just to frame as many walls as your deck or slab will accommodate (figure that you can frame an 8-ft. wall if you've got at least 100 in. of room on the deck for its height), and then raise them, repeating this procedure until there aren't any sets of plates left.

The layout will have a lot to say about which walls get framed first. Exterior walls get priority, so you can save the precious working room near the center of the deck as long as possible. Of the exterior walls, you will be framing the by-walls first and then the butt-walls so that you can build as many walls in place as possible. Pick one of the longest exterior by-walls to begin; the back wall of the house is the traditional place to start.

Studs and plates—The best place for a lumber drop is right next to the slab or deck. This way you can literally grab a stud when you need one. But on a steep site or second story you'll need to pack the lumber to the deck as

you need it. A laborer will help in this situation, but don't give in to the temptation to stockpile studs on the deck—you'll only end up having to move them again. You can keep plate stock handy, though, by leaning it up against first-story framing. Spot two or three bunches of 10 to 20-footers around the building for double top plates. It's pretty easy to take the bows and crooks out of a double top plate when your'e nailing it, but if it's real bad, cull it out and start a pile that you can cut up into blocking and short jacks.

Nails—The only other items on the deck should be a skillsaw and a 50-lb. box of nails. The size and kind of nails a framer chooses seem largely regional. The allowable minimal size depends on whether you are *face-nailing* (through the face of one board into the face of another, such as nailing down the double top plate), *end-nailing* (through a board face into the end of board, such as nailing through the bottom plate into the studs) or *edge-nailing* (through the face of a board into the edge of another, such as a channel)—see the drawing, facing page, top. But rather than carry 8d, 10d, 12d, and 16d nails, it's easier just to carry a handful of eights in the small pocket of your nail bags for 1x let-in bracing and toe-nailing, and sixteens in the big pocket for everything else.

There are lots of choices when it comes to

nail coatings. Brights (regular steel nails without any coating) and even galvanized nails are okay, but I like *sinkers*. Sixteen-penny sinkers are a hybrid nail made in heaven (actually in Asia) for framers. Because of their coating, it takes half as many swings to drive one as an uncoated nail, and they don't crumple when you take a healthy swat. Their shank diameter is larger than a box nail, but not as thick as a common, which will split the ends of dry plate stock. They are also slightly shorter than 3½ in. (the length of the usual 16d nail), and have a thicker head. Sinkers can be either cement coated or vinyl coated.

In much of the far West, green vinyl (sold as g.v.) sinkers have become the predominant framing nail in the last five years. Although the vinyl does reduce the friction when they are being driven, these nails don't seem to offer much resistance on their way back out compared with other varieties. They're not nearly as bad, however, as nails that have had a treatment tract piece-workers affectionately call "gas 'n' wax." This is a coating made from kerosene and beeswax that is applied on site away from the eyes of building inspectors. You can whisper to these nails and they will drive themselves, but they unfortunately withdraw with the same ease.

My favorite nails are cement-coated (sold as c.c.) sinkers. The resin coating heats up from the friction of entering the wood and

Stocking the wall. Some framers stock the wall with studs before they split the plates apart (facing page), so they know how far back to carry the top plate. But separating the top plate and stocking it with headers first will define the openings right away and save having to flop heavy headers down in a sea of 2x4s. Either way, crowning the studs pays off.

makes the nail slippery. But unlike the green vinyl sinkers, once the nail is in place, the coating bonds with the surrounding wood, and holding power increases several times over brights. The disadvantage of c.c. sinkers is that the black resin accumulates on your fingers, nailbags and hammerhead. If you're nailing finish boards with these sinkers (such as 2x tongue-and-groove roof decking over exposed beams), coat your hands with talcum powder before starting to keep from leaving black fingerprints all over the ceiling.

Building the walls—Unlike Jud Peake, who details plates flat in his system (see his article on pp. 32-40), I'm used to detailing plates up on edge. To frame a wall in place, I take the tacked-together plates and lay them just inside the snapped layout (wall) line with their inside edges down. Then I drive a 16d toenail through the bottom face of the bottom plate into the deck every 10 ft. or so. This way, when the top plate is separated, the bottom plate is already in position for framing. And once framed, the wall can be raised in place as if it were hinged to the floor.

Using this system, I separate the top plate with my hammer claw, walk it back on the deck and stock it with its headers first thing. This sets the location of wall openings early so you don't stud over one by mistake. It also means you don't have to flop a long, heavy header down in a sea of studs. But lots of other framers like to stock the wall with studs before separating the top plate so they know how far back to lay it (photo facing page).

Whichever order you use to stock the studs, cull the ones that are really crooked, and make sure that you lay all of the keepers crown up. Do this by sighting along each one before laying it down. Once you've got a stud on every layout mark, you can add corners and channels if you've made them up as units. Make doubly sure that the flat stud in each channel is facing as it should by looking at the layout on the floor as well as at the marks on the plate.

So far you haven't driven a nail, but soon that's all you'll be doing. If I'm using 4x12 headers and don't have to deal with head jacks, I like to nail the top plate to the top edge of the header first thing. This adds a lot of weight to the top plate and keeps it from moving around. Also. top-plate splices often come in the middle of a header, and you can begin making the wall a single unit by connecting the top plate to the header. Make sure the plates butt tightly so you don't lengthen the wall, and drive two nails into each plate end. At each end of the header, you'll also need two sixteens. In between, you should

Illustrations: Christopher Clapp

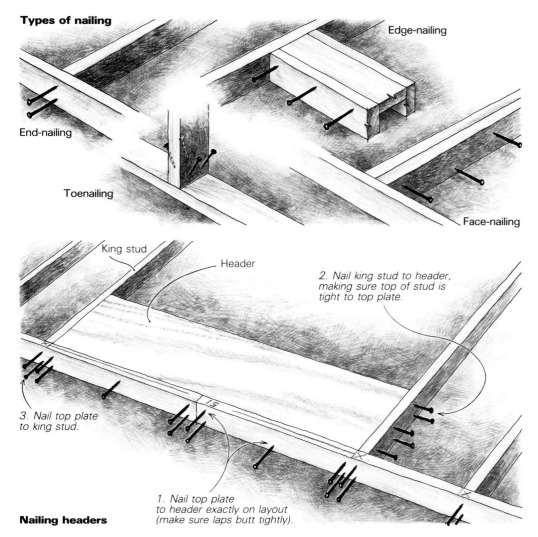

Types of nailing

Edge-nailing

End-nailing

Toenailing

Face-nailing

King stud

Header

2. Nail king stud to header, making sure top of stud is tight to top plate.

3. Nail top plate to king stud.

1. Nail top plate to header exactly on layout (make sure laps butt tightly).

Nailing headers

stagger the nailing to each side at 16-in. centers. If the plate runs through without a splice, use two 16d nails at each end of the header, and then stagger nails every 16 in. in between, as shown in the drawing above).

Next, you should take care of the king stud to make sure that you have room to swing your hammer. Drive at least four nails through the face of each king stud into the end of the header. With a 4x12 header I use six 16ds. This is an important intersection. If the king stud doesn't stay tight to the header, it will pop sheetrock nails and leave a crack radiating away from the corner of door and window openings. Last, end-nail the top plate to the king stud with two nails. If you are using a header that doesn't reach the top plate and therefore requires head jacks, you can drive two nails into the top of each one through the top plate as you would a regular stud, and then toenail the bottoms to the top of the header with four 8d toenails each.

Now you can start nailing off everything that you've laid between the plates. Stay with the top plate and begin nailing at one end or the other. If you are right-handed, you'll find that working from left to right will probably be most comfortable and help you establish a rhythm. You'll be working bent over from the waist—one foot up on the edge of the plate, and the other foot nudging the stud onto the layout line and bracing it from twisting (photo

next page). Each 2x4, whether it's a layout stud or part of a corner or channel, gets two 16d nails driven into it through the plate.

Your first nail should be near the top edge of the plate, where the pencil layout line is marked. Set the nail with a tap of your hammer, then line the stud up on the mark and drive the nail through the plate and into the stud with your next couple of blows. Be careful to split the difference if the stud is narrower or wider than the plate to which you are nailing it. This may mean having to hold the stud up with your nail hand until you get the first nail in. Pay close attention to the layout; it's surprisingly easy to lose your concentration and begin nailing on the wrong side of the line despite the X on the edge of the plate that indicates the stud location. The only trick to the second nail is to make sure the stud is square to the plate. Judge this by eye.

Production framing requires a strong hammer arm and a dexterous nail hand. The only way to develop your arm is to drive a lot of nails, but there are some tricks to fingering the nails. With 16d nails, you need to orient the heads all in the same direction. I like to do this when I'm over at the nail box to refill my bags. This way, each time that you reach down you can pull out a large handful of nails ready to drive. Then, without dropping the nails cradled in your hand, use your index finger and thumb to reach into your palm to

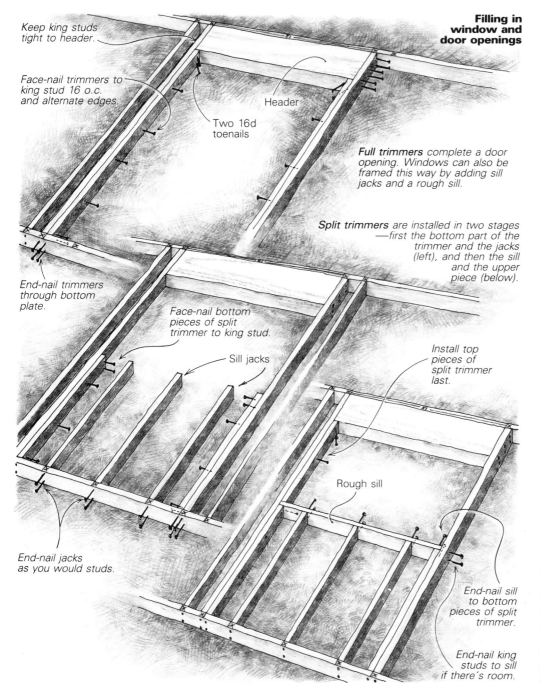

Keep king studs tight to header.

Face-nail trimmers to king stud 16 o.c. and alternate edges.

Header

Two 16d toenails

Full trimmers complete a door opening. Windows can also be framed this way by adding sill jacks and a rough sill.

Split trimmers are installed in two stages —first the bottom part of the trimmer and the jacks (left), and then the sill and the upper piece (below).

End-nail trimmers through bottom plate.

Face-nail bottom pieces of split trimmer to king stud.

Sill jacks

Install top pieces of split trimmer last.

End-nail jacks as you would studs.

Rough sill

End-nail sill to bottom pieces of split trimmer.

End-nail king studs to sill if there's room.

pinch a single nail. Extend the nail and rest it point down on the plate for the hammer while your other fingers regrip the nails in your palm. This should all happen while you are backswinging your hammer.

The first swing should be just a tap to start the nail. All carpenters hit their fingers occasionally, but you learn to keep your fingers out of the way when you are swinging hard enough to do any damage. This is particularly important with framing hammers, which can tear as well as bruise.

Once you've finished nailing off the top plate, move to the bottom plate and do the same thing all over again. You may want to reverse this process if you're framing exterior walls on a slab. The big problem here is that the anchor bolts invariably fall on the stud layout, and this requires chopping out some of the stud bottom. This is a lot easier if the top of the stud isn't already nailed.

When you are nailing the end studs on walls that butt by-walls on both ends, hold this last stud back from the end of both plates about ¼ in. This way if the stud bows out slightly, it won't prevent the wall from being raised. You can drive it back to its proper position when you nail the intersection together.

Another potential trouble spot is plate splices. If they come at a stud, you will need four 16d nails in the end of that stud—two from each plate end (photo below left). The other place that plates often join is in a doorway. Although the bottom plates will eventually be cut out, on long, heavy walls make sure that they stay butted by nailing a block on top of the plates at the joint.

Windows and doors—Once all of the headers, studs, corners, and channels are nailed in, you can complete the openings. Doors are easy. Fill in under each header with one trimmer on each side (two on each side if the opening is 8 ft. wide or more), as shown in the drawing at left. Trimmers need to fit snugly between the header and the bottom plate. You shouldn't have to pound the plate apart to get them in, nor should they be short. At its bottom, nail the trimmer like a stud. Then face-nail it to the king stud at 16 in. o. c., with the nails alternating from edge to edge. At its top, drive two 16s up at an angle to catch both the king stud and the corner of the header.

Taking some care with the trimmers will really pay off when it comes time to hang and case the doors. Make sure that the bottom of the trimmer is nailed right on its line—this will ensure a plumb and square opening. Also make sure that the edges of the trimmer and the king stud are even—this means that the

Framing. **The top plate being nailed at left is a splice over a stud, which requires four nails driven on a slight angle. When space gets tight on the deck, two framers often end up working on the same wall. One should take the entire top plate; the other, the bottom plate. If they are both right-handed, they will begin on opposite ends of the wall since they will be moving left to right for comfort and speed.**

mitered door casing will sit on a flat surface instead of a hump or dip. When you're finished nailing the trimmers, go back to the top of the king stud with your hammer and give it a bash or two so that it isn't separated at all from the end of the header.

Windows are a bit more complicated than doors because you've got the rough sill to contend with. Here you've got a choice. The sill can be cut to butt against both king studs, with *split trimmers* nailed up above and below it, or you can nail up full trimmers and toenail the sill into them. I was taught to use the split-trimmer method, and I prefer it because you can end-nail the sill into the lower trimmers and even end-nail through the king stud for a very solid connection.

If you're using split trimmers, begin by cutting the rough sill to length. Instead of measuring between the king studs, hold your sill stock up to the bottom of the header and mark it. Here the framing will be tight, so you'll get an accurate measurement even if the studs bow out down where the sill will be installed. Next, cut the sill jacks if their length is given in the layout, nail them at the bottom plate and drop the rough sill on top. If the sill-jack measurement isn't given, the size of the rough opening will be, and you can use your tape measure to mark the top of the rough sill. Allow 1½ in. for the sill, strike a line, and use this lower mark to cut in the sill jacks.

Remember that the lower part of a split trimmer is just another sill jack except that it also gets face-nailed to the king stud using 16d nails on alternating edges at 16 in. o. c. (drawing, facing page). Once all the jacks and lower trimmers are installed, you can end-nail the rough sill to them with two 16d nails at each intersection. If there's room, drive a couple of end nails through the outside of the king studs into the sill. If you can't do this, drive a 16d nail at a slight angle down through each end of the sill to catch the king stud.

The last step is to cut the top half of the split trimmer and face-nail it in. The only other complications are double sills and trimmers for very wide windows. In this case, install the first trimmer on each side full, and split the inside one around the doubled sill.

If you're not using split trimmers, you can treat the opening as you would a door at first by installing the trimmers full length underneath the header. Then mark the sill for length, and cut it. Follow the same procedure that was outlined above to set the sill and its jacks. But when you set the sill, use at least four toenails from the sill into the trimmer, along with a 16d toenail through the thickness of the trimmer into the end grain of the sill.

Rake walls and specials—Rake-wall (gable-end wall) framing is a little different because the tops of the studs have to be cut at the roof-pitch angle. But if the wall has been laid out right, cutting shouldn't slow you down too much. Mark the studs in place and cut their tops with the shoe of the saw set at the roof-pitch angle.

The bottom plate nails to the stud normally.

Blocking. **Horizontal blocking for tubs and showers on interior walls goes faster if you let it in rather than cutting and nailing short blocks in the stud bays. Make sure to set your saw depth accurately to ensure that the blocking ends up in the same plane as the rest of the wall.**

At the top plate, I usually drive a toenail through the toe (long point) of the angle at the top of the stud into the plate to hold the stud on the line. Then I drive two 16s down through the top plate into the stud, as you would in a standard wall.

The last step before double top plating any wall is to take care of the specials, which usually means different kinds of blocking. Interior soffits (drop ceilings) and walls over 10 ft. high will need fire blocking or stops—horizontal 2x4s nailed between studs to delay the spread of fire up the wall. Chalk a line across the studs and stagger the blocks on either side of the line with two 16d nails driven through the studs into each end of the block.

Mid-height flat blocking may also be required on exterior walls that will be covered with stucco. Production framers often raise this blocking a few inches on 8-ft. walls to 51 in. or 52 in. off the floor. This makes it easier to duck through the stud bays when working, more convenient as a shoulder support for lifting the wall onto foundation bolts on a slab, and keeps the blocks just slightly higher than electrical switch boxes.

Blocking for tubs and showers is a bit easier since it is usually installed with the face of the blocking nailed flush with the edges of the studs. It can either be cut in short blocks (14⁷⁄₁₆ in. for 16-in. centers), or a length of 2x can be let into the studs (you can do this only on interior walls where the side that requires blocking is facing up). To let in blocking (photo above), position the 2x4 on the wall where the blocking will be needed and scribe

Bracing and raising. Let-in bracing can be cut into the top edges of the studs by setting the saw depth to 1½ in. and letting the shoe ride on the bracing (photo right). This procedure is not without risk of kickback. The safer alternative is scribing against the brace and cutting to the pencil mark. Facing page: The let-in brace in the foreground of the photo shows a forest of 8d nails that will be driven once the wall is raised and plumbed. Two keys to raising walls safely are even distribution of the weight and vocal coordination of effort.

Let-in bracing

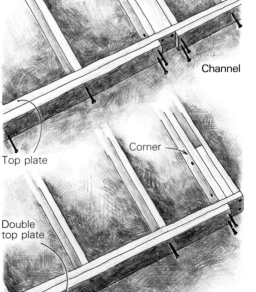

Use three 8d nails at bottom plate.

Bottom plate

1x4

Vertical shoulder cut

Drive nails home only at bottom plate and first stud so wall can be racked once it's vertical.

Double top plating

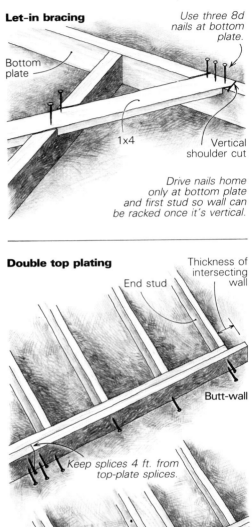

Thickness of intersecting wall

End stud

Butt-wall

Keep splices 4 ft. from top-plate splices.

Channel

Top plate

Corner

Double top plate

By-wall

a pencil line on either side. Cut to the inside of each of these lines with your saw set at 1½ in., remove the resulting scrap and nail the 2x4 in place.

Double top plate—The key to double top plating is the channels and corners. When two walls intersect, one of the double top plates acts as a tie between them. The double top plate on a butt wall will overhang its end stud 3½ in. for a 2x4 wall, and the double top plate of the corresponding by-wall will be held back 3½ in. to receive it. Double-plate splices should be held back at least 4 ft. from the end of a wall or from splices in the top plate.

Pull your plate stock up on the deck and lay it down where it will go. You may be tempted to use a tape measure to mark the overhang or hold-back on butt-walls. Instead, hold the stock to the far side of the channel and scribe the other end of the stock with your pencil held against the end of the wall. As long as the walls are all the same thickness, you will get a double top plate that's the right length by cutting at the pencil line.

Double top plating can go very fast. The ends of each piece require two 16d nails (drawing, left). In between you'll want to drive a 16d at every stud, alternating sides of the plate. Hitting the stud layout with these nails allows you to let in bracing in any of the bays without worrying about hitting a nail with your saw. You will only get to nail double top plates on butt-walls that are built out of place, since this plate has to project beyond its end stud by the thickness of the by-wall it will intersect. If it's easier to build the wall in

place, cut the plate anyway and nail it once the wall is raised.

Accuracy in cutting and nailing double top plates is essential. A double top plate that projects a little beyond a corner will drive you bananas when you're out on the scaffolding later nailing the sheathing. Be careful with the width of channels too—as soon as you leave a 3½-in. slot for the double top plate of an intersecting wall, it will invariably be cut from dripping wet stock and measure 3⅝ in. You can trim the double top plates on each side to make it work as it should, but you'll have to lay the wall back down to do it.

There's another precaution you should take just before raising a butt-wall that has been framed out of position. Give the overhanging double top plate a couple of blows from your hammer to drive it up off the top plate a half inch or so. This way when you are sliding the butt wall into place, the projecting double top plate won't hang up on the top plate of the already standing by-wall.

Bracing—All walls need some kind of corner bracing to prevent them from racking. There are lots of ways to get this triangulation. Sheathing and finished plywood siding provide excellent resistance if the nailing is sufficiently close. Cut-in bracing (flat blocks cut at an angle and nailed into each stud bay along a diagonal line) and metal X-bracing (long 16-ga. sheet-metal straps that are nailed to the studs under siding) are both quite effective. But for maximum strength when you are not using plywood, let-in bracing is usually specified. This 1x4 brace is mortised flat into

the exterior edges of the studs. It should extend from near the top corner of the wall down to the bottom plate at about 45°. In an 8-ft. wall with studs at 16 in., the brace will cut across six stud bays. You can get one into five bays by increasing the angle a bit.

Not every stud space needs to have a brace; the minimum standard for exterior walls and main cross stud partitions is one brace at each corner, and one at least every 25 lineal feet in between. But the more braces you use, the easier the wall will be to plumb and line, which in turn will create square rooms with plumb door jambs. And each wall that gets a let-in brace will act as a single unit, which in turn will increase the strength of your frame when it's subjected to wind or seismic loads. On a quality house, the small price you pay for 1x4 and the couple of minutes it takes to let it in allow you to brace even short walls. This goes for interior partitions as well. I even like to shear panel (⅜-in. plywood with close nailing) long cross walls.

A typical medium-length wall (say 30 ft. long) contains two let-in braces. These should form a V with an open bottom, since the tops of the braces start high on each corner. Of course it's not always this simple, because most walls contain window and door openings that have to be dodged.

These braces are best let in before the walls are raised, but you must make sure that the wall is close to being square. If you're framing the wall in place on the deck, it should be okay. But if it's been moved around some and is visibly racked, then you need to rough-square it by getting the diagonal measurements approximately the same. Then lay a 1x4 across the top edges of the studs at an approximate 45° angle. Both ends should overlap the plates between studs. On an 8-ft. wall this will require a 12-footer. Now scribe each side of the 1x4 on every stud and at the top and bottom plates. Cut on the outside of each of these lines with your saw set at a fat ¾-in. depth. This will produce a slot about ¼ in. wider (two saw kerfs) than the brace—the extra width will accommodate the racking of the wall left and right during plumbing and lining. Remember, it's the number of nails and their shear strength that will keep the wall square eventually, not the fit of the brace in its slot.

Production framers eliminate scribing by cutting with the 1x4 in place (photo facing page). This procedure works best with a worm-drive saw, and requires a lot of experience with the saw because of the danger of kickback. Even then, safety takes a back seat to speed here. The brace is held down in place with one foot, and the saw is run along one side of it set to a depth of 1½ in. The shoe of the saw rides right on the brace. The framer will then change directions and saw back along the other side of the 1x4. The ends of the 1x4 are cut off in place at the bottom of the bottom plate and about ¼ in. shy of the top of the double top plate. The last step is to chop out the little blocks of wood between the saw kerfs.

Install the brace while the wall is still flat on the deck. Drop the brace into its slot, holding it flush with the bottom of the bottom plate, and drive three 8d nails there. You can also nail it to the first stud since this is still low enough that it won't interfere with racking the top of the wall during plumbing and lining. You should also start two nails in the face of the 1x4 for each stud so that they can be easily driven home with one hand later. Start a total of five nails at the top of the brace—two into the double top plate and three just below this at the top plate. A lot of framers also start a nail in the top plate and bend it over the brace to keep it from flopping around when the wall is being racked.

Let-ins are most effective if they have a vertical shoulder at the bottom plate rather than coming to a point. This means making a plumb cut on the last inch or so of the brace measured along the angle, and a corresponding slot in the bottom plate (top drawing, facing page). Not all building inspectors will insist on this, but it's a good idea anyway.

If you're building on a slab, don't allow the let-in brace to sit on the concrete. End the brace in the middle of the 1½-in. thickness of the redwood or pressure-treated bottom plate.

Raising the walls—This is the best part, but also the point at which a lot of backs get wrecked (for the mechanical alternative, see sidebar, p. 48). Move anything you might trip over. If the wall is a rake, or is very tall or heavy, toenail the bottom plate to the wall line, or nail lumber strapping to the bottom of the plate, run it under the wall, and nail to the subfloor on the other side. For a standard 8-ft. wall, nail stops—short lengths of 2x4—to the joist just below the deck every 6 ft. to 8 ft. so that they stick up where the wall is going to be raised, preventing it from skidding over the

Mechanical wall raising

Every carpenter has at least one story about a former partner or laborer who was 6 ft. 4 in. and immensely strong. However, being able to lift twice your body weight is of little consequence when you consider the weight of construction materials and the power of machines. While hiring a crane or forklift is usually not an economical option on small jobs, using simple mechanical advantage is, and it can change the way you work.

Wall jacks are a good example. With a pair of them, two people can easily raise a long wall full of solid headers weighing 2,000 lb. From there, the logical step is to sheathe exterior walls—and even add windows, siding, trim and paint—down on the deck before they are raised.

There are two kinds of wall jacks, but they work similarly. The first looks and operates much like a scaffolding pump jack. By pumping the handle of its ratchet winch (the same mechanism a car jack uses), the jack walks up a 2x4 (or 4x4), carrying the top of the wall to be raised with it on its horizontal bracket. In fact, these devices are often called walking jacks. The 2x4, which begins in a vertical position, is held from skidding by a block nailed to the deck. As the jack makes its way up, the 2x4 is allowed to get less and less vertical so that the wall will continue to bear on it. These jacks are relatively inexpensive, ranging from $60 to $125 apiece. Two brands I know of are Hoitsma walking jacks (Box 595, River Street Station, Paterson, N. J. 07524) and Olympic Hi-jacks (Olympic Foundry, Box 80187, Seattle, Wash. 98108).

The other kind of wall jack (the one that I've owned) is manufactured by Proctor (Proctor Products Co., 210 8th St. South, P. O. Box F, Kirkland, Wash. 98033). It consists of a metal boom that is fitted to a hinged plate at the bottom that nails down to the subfloor and joists below. What amounts to a ¾-ton come-along is mounted on the boom just below waist height. The 5/32-in. galvanized aircraft cable is threaded through a sheave (pulley) at the end of the boom and is fitted with a nailing bracket that attaches to the double top plate of the wall to be raised. The boom begins in a vertical position, and begins to lean as soon as the ratchet winch is put to use and the wall begins to come off the deck. An

adjustable stop prevents the wall from going beyond vertical. Proctor wall jacks come in three lengths (16 ft., 20 ft. and 23 ft.) and all of them telescope for carrying or storage. Although this system is more expensive— the smallest pair of jacks retails for $445— it's very safe and can handle walls as long as 75 ft. with just two carpenters on the job.

The real advantage to wall jacks is that they give you the freedom to finish a wall completely while it's flat on the ground. This means money saved even if you're building a one-story house on a flat lot. On high work, you often can eliminate the cost of scaffolding and speed the usually slow progress of sheathing, windows, siding, trim and paint by doing most of the work right where the frame sits on the deck. On the houses where I've chosen to do this, I've also been able to use less skilled labor to complete the walls. For instance, any careful person can apply a full-bodied stain with a roller when the wall is flat and accessible. It doesn't have to be sprayed by a painting sub. Trimming out windows can be left to an apprentice, since having to recut a piece for fit isn't a big deal when you're not climbing down off scaffolding to do it.

Tilting up finished walls isn't for every house. It is best used on modest rectangular plans with long walls where the siding and the trim detailing are simple. In any case, corners have to be finished off from ladders or simple wood scaffolding suspended from nearby window sills.

The only real trick to completing a wall on the deck is to rack the framing so it's exactly square before you sheathe it. This is worth checking several times. The alternative is spending half an hour with a cat's paw removing all that shear nailing so that the ends of the wall will sit plumb once in place. To check for square, toenail the bottom of the bottom plate of the wall to the deck right on the snapped wall line and then use a sledgehammer to bump the top of the wall until the diagonals measure the same. The only other complication is with butt-walls once the by-walls are up. Butt-walls have to be finished slightly out of position (left and right) to buy clearance for the sheathing or siding that will overlap the corner framing of the by-wall, or the end panel will have to be left off until the wall is raised. —P. S.

edge. Also nail a long 2x4 with one 16d nail high up on each end of a by-wall. The single nail will act as a pivot so that the bottom of the brace will swing down and can be angled back to about 45° alongside the deck. It can then be nailed to the rim joists when the wall has been raised to approximately plumb. More brace material should be stacked nearby so that it can be grabbed quickly.

To get the wall in a position where you can get your hands under it, lean short lengths of 2x against the face of the double top plate every 12 ft. or so. Then, standing inside the wall itself, bury your hammer claw in the double top plate with a healthy swing. When you lift the top of the wall just a few inches, the blocks will fall beneath the top plate and you'll have enough room to get a grip.

Now it's time to gather your crew. Most carpenters can lift a good 12 lineal feet of 2x4 framing—more if there aren't a lot of 4x12 headers, less if the wall is 2x6 or framed with very wet lumber. Spread people out along the wall according to where the weight is. Headers are the worst because the weight is all at the top. The ends of the wall are almost always the lightest. The first maneuver in lifting the wall (photo previous page) is called a clean and jerk in weightlifting. If you don't bend your legs it's a sure road to a hernia or a bad back. The second stage—where you've got the wall to your waist and you are pushing with the palms of your hands—is basically a press and should be done with your legs braced behind you. Don't make it a contest. Raising a heavy wall requires staying in sync with everyone else.

If you're raising a by-wall, at least one of the crew should let go once it's up—with fair warning—and nail the outside braces. The wall should lean out slightly at the top to leave a little extra room for the butt wall that will intersect it.

Raising walls on a slab is slightly different. Here you've first got to raise the wall to a vertical position, and then lift and thread the bottom plate onto the foundation bolts. Long 2x4s on edge can be used effectively as levers under the bottom plate to lift the wall up above the bolts, as long as you have someone steadying the top of the wall. End braces can be nailed off to stakes driven in the ground right next to the slab. Once the wall is steady, beat on the sill at several spots to make sure that it's down, then put washers and nuts on the foundation bolts and screw them down finger tight.

If you're raising an exterior butt wall on either a slab or a deck, you'll be nailing the end stud to the corner of its corresponding by-wall, rather than using a brace. Make sure that the bottom plate is flat on the deck and that the two top plates match in height. Also align the outside face of the corner and the edge of the end stud so that they are in the

same plane all the way up. Alternate 16d nails on each side of the end stud every 16 in. If you are raising a partition or interior wall, nail its end stud to the channel in the same way, with the same kind of care.

Long walls may require an intermediate brace or two before everyone can let go. How much bracing you need to add depends in part on how soon you'll be going home for the day. Braces take up a lot of space on the deck when you're trying to frame the rest of the walls. But when you leave the site, it's a whole different story. Figure that a hurricane will strike that night and brace everything off accordingly, especially if you have already sheathed your walls.

If you're bracing off walls on a concrete slab for the night or weekend, you can use a "tepee" on exterior walls. To make one, take a 14-ft. piece of plate stock and run it through a stud bay of the wall to be braced so that half

of the plate stock is cantilevered out beyond the building, and the other half is on the inside of the slab. Then nail a 2x4 brace from the top of the wall down to the end of the plate stock on both the inside and outside. This kind of brace allows a little give, but the wall won't go anywhere.

A precaution you want to take on a plywood deck is to nail down the bottom plate. First make sure that the ends of the wall are where you want them—on a layout mark, flush with the perimeter of the deck, or butting another wall. Then use a sledgehammer to persuade the bottom plate into a straight line that sits just at the edge of the wall line established in chalk during layout. After that, drive one 16d nail per bay to keep it there.

The last thing that you have to do to connect the walls is to nail down the double top plates that lap over intersecting walls. Remember that the walls don't have to be

plumb; you'll take care of that later. You are just making sure that the walls are nailed together exactly as they were laid out from the bottom plate all the way up.

To nail off the double top plates that lap, you can claw your way up a corner and walk the plates between channels. But you'll probably get it done more safely by moving a 6-ft. stepladder around. First, make sure that the walls are driven together tightly all the way up, and that they are aligned with each other vertically. Then you can finish off the double top-plate nailing using four 16d nails for each 2x4 plate lap.

When all the plate pairs are nailed down, your frame will be complete. Plumbing and lining it will make it ready for joists or roof rafters. The feeling at the end of a day of this kind of work is unparalleled. You are surrounded by the tangible evidence of your progress and the worth of your labor. □

Plumb and Line

Without this final step of straightening the walls, the care taken during framing will have little effect

by Don Dunkley

It's strange to think of a completed wall frame as being a kind of sculpture that needs final shaping. But that's just what it is. Until the walls have been braced straight and plumb, they can't be sheathed or fitted with joists or rafters without producing crooked hallways, bowed walls, ill-fitting doors and roller-coaster roofs.

The production name for getting the frame plumb, square and straight is *plumb and line.* The job doesn't take very long—three to six hours for most houses—but it's essential. After the frenzied pace of wall framing, plumb and line can be a welcome relief. It requires at least two carpenters (three's a luxury) working closely together. The work is exacting, but not hard, and there's a sense of casual celebration in having finished off the wall framing.

Stud-wall framing is based on things being parallel and repetitive (see article on pp. 32-40). If you plumb up the end of a wall, then all the vertical members in the wall will be plumb in that direction. And if the bottom plate of the wall is nailed in a straight line to the floor, then getting the top of the wall parallel to the bottom is easy: just plumb the face of the wall at both ends, and make the top plates conform to a line between these two points.

Plumb and line is a fluid process in which walls are braced individually, but the sequence of operations is important. Although you can start at any outside corner, walls should be plumbed and then lined in either a clockwise or a counterclockwise order, since each correction will affect the next wall. Once the bottom plates have been fully nailed to the floor or bolted and pinned to the slab, the exterior walls are plumbed up and the let-in braces are nailed off, or temporary diagonal braces are installed to prevent the wall from racking (the movement in a wall that changes it from a rectangle to a parallelogram, throwing the vertical members out of plumb). Then line braces are nailed to the walls to push or pull them into line at the top and to hold them there. Last, the interior walls are plumbed and lined with shorter 2x4 braces.

Plumbing and lining can begin once all the intersecting walls are well nailed to their channels or corners. On each of these walls,

Holding wall intersections tight while the lapping double top plate is nailed off can make the difference between a plumb frame and having to make a lot of compromises later on.

the end stud has to line up perfectly with the channel flat or corner studs. It's also wise to make sure that the heights of the walls match up. If they don't, it's usually because the end stud isn't sitting down on the bottom plate. Also, all double top plates must be nailed off where they lap at corners and channels. Be sure that there aren't any gaps here (a common defect in hastily framed houses), and that the end studs and plates are sucked up tight (photo facing page).

Fixing the bottom plates—The next job is to toss all the scraps of wood off the slab or deck and sweep up. This way you'll be able to read the chalklines on the floor. If the frame is sitting on a wood subfloor, the bottom plates have likely been nailed off. On a slab, though, you'll have to begin by walking the perimeter of the building putting washers and nuts on all the foundation bolts. Tighten the nuts a few turns before using a small hand sledge to knock the plates into alignment with their layout snaplines. The slight compression created by the nut will keep the plate from bouncing when you hit it.

Getting the plates right on the layout is critical, because the bottom of the frame determines the final position of the top of the wall. Once the plates are where you want them, run the nuts down tight. The fastest tool for doing this is an impact wrench, found mostly on tracts and commercial jobs. Next best is a ratchet and socket, with last place going to the ubiquitous adjustable wrench.

Wherever bottom plates butt end-to-end on a slab, both pieces must be fastened down with a shot pin. There are many powder-actuated fastening devices currently available; the gun I like the best is the Hilti. Unlike many of the older guns, which have space-hogging spall shields, it is slim enough to get into tight spots between studs. The model that I've used (DX-350) is fitted with a multi-load magazine, so it can be reloaded in seconds. It also uses a pin and washer that are made as a single unit—a big advantage.

If there are two of you working, one person can set the pins into the top of the plate with a tap of a hammer. The other person can follow, slipping the gun barrel over the shaft of the pin and firing. The pins should be placed close to the inside edge of the plate so that you don't blow out the side of the slab. To prevent the bottom plate from splitting out at the ends, you can nail a ⅜-in. or ½-in. plywood scrap over it and then fire the pin through it. Since doorways take a lot of abuse, a pin should be shot into the sill at each side of the opening.

After all the outside walls are fastened down, you can line the bottom plates of the inside walls and shoot or nail them down. Be especially careful on long parallel runs like hallways—any deviation will really shout out when the drywall is hung and the baseboard installed. On interior partitions, one drive pin or nail every 32 in. should be sufficient. Make sure to hit the flanking bays of doorways, and both sides of bottom-plate breaks.

From *Fine Homebuilding* magazine (October 1984) 23:68-71

Setting up—Once all the bottom plates are secure, spread your bracing lumber neatly throughout the house. Interior walls will need 8-ft. and 10-ft. 2x4s. Exterior walls will need 14 and 16-footers for line braces, and shorter diagonal braces if the wall doesn't have let-ins. You'll have to guess how much bracing you'll need. More is always better than less, but figure that you'll need a line brace every 10 ft. and a diagonal brace on any wall over 8 ft. long without a let-in.

To plumb walls I use an 8-ft. level with two small aluminum spacer blocks that fit over the lip of the level near each end, and are tightened with handscrews. When the level is held vertically against a wall, the spacers butt the bottom plate and double top plate and prevent bowed studs from interfering. You don't need to buy an 8-footer; any level that is dead-on accurate will do if you extend it to the length you need (photo right). Select the straightest 2x4 around, cut it to 8 ft., and tack a short piece of 1x to one edge of the 2x4 at each end for a spacer. Then attach the level to the other edge of the 2x4 using duct tape or bent-over 8d nails.

Plumbing the walls—Exterior walls are plumbed from their corners. Choose any one as a starting point. Position the level at the end of the wall so that you are measuring the degree to which it's been racked. The bubble will indicate which way you have to rack the wall back to plumb. To do this, you can make a push stick out of a 9-ft. to 10-ft. 2x4 that is largely free of knots. A more flexible push stick can be made from two 1x4s nailed together face-to-face. Cut a slight bevel on the bottom end to keep it from slipping. To use the push stick, place the top of it in a bay about midway down the length of the wall to be racked, up where the stud butts the top plate. The stick should angle back in the opposite direction from which the top of the wall should be moved. Plant the bottom end on the slab or deck right next to the bottom plate. Pushing down on the center of the stick will flex it, exerting pressure on the top of the wall. Keep pushing and the wall will creak into a plumb position.

Racking walls takes coordinated effort. If you're on the smart end of the things (using the level), you need to shout out which way the wall must move and how far. Although overcompensation is usually a problem at first, soon your partner (who's on the push stick) will get used to what you mean by "just a touch" and be able to make the bubble straighten up as if he were looking at it.

Once you've racked the wall to plumb, hold it there permanently by nailing off the let-in bracing (photo right). But before you do that, have your partner keep the same tension on the push stick while you plumb the other end of the wall as a check. If it's plumb also, nail off the braces on the wall as fast as you can. Here it's handy to have a third person to do the nailing while you keep your eye on the bubble. If for any reason (poor plating or a gap where the top plates butt) the wall isn't

Plumbing. The length of the level you use is a lot less important than being sure that it is dead-on accurate. Here a 4-ft. level is attached to a straight stud with bent-over nails. Using 1x spacers at the top and bottom of the stud (or manufactured aluminum spacer blocks on an 8-ft. level) will ensure an accurate reading despite bowed wall studs.

Keeping it plumb. While the let-in brace that will keep the wall plumb is being nailed off, this short wall is being held plumb by tension on a push stick made of face-nailed 1x4s.

plumb on the opposite end, you'll have to split the difference.

Walls that are shorter than 8 ft., walls with lots of doors and windows, and exterior walls that will get shear panel, sheathing or finished exterior plywood won't have let-in bracing. For these walls, a temporary 2x4 brace must be nailed up flat to the inside of the wall at about 45°. Be sure that none of these braces extends above the top plate. Use two nails at the top plate, one nail each into the edges of at least three studs and two nails into the bottom plate. As with all plumb-and-line bracing, drive the nails home (you won't be pulling these braces until joists, rafters and sometimes sheathing are in place). Continue this process until all exterior walls are plumbed.

Lining—The next step is to get the tops of the exterior walls straight. If you've got a practiced eye, you can get the best line of sight on a wall by standing on a 6-ft. stepladder with your eye right down on the outside edge of the double top plate (photo above). This is the way I usually line, but for absolute accuracy you can't beat a dryline.

Set this string up by tacking a 1x4 flat to the outside edges of the top plates at each end of the wall. Then stretch nylon string from one end of the wall to the other across the faces of the 1x4s, and pull it tight (a twist knot looped and tightened on a nail will hold a taut line). The 1xs hold the line away from the wall so that if the wall bows, the string won't be affected. To determine whether the wall is in or out at any given point, use a third scrap of 1x as a gauge between the string and the top plates. If an inside wall intersects with the outside wall, I plumb the end of the interior wall, and then check the rest of the wall with the 1x gauge.

To make corrections in the wall and then hold it there, use line braces. These are 2x4s on edge that are nailed high on the wall and angle down to the floor or the base of an interior wall. To be most effective, they should be perpendicular to the plane of the wall. If a wall is at least 8 ft. long, it gets a brace—even if it's already straight. Braces not only correct the walls, but also make them secure enough to walk on while laying out and nailing joists or rafters, and keep these ceiling members from pushing the walls outward.

Usually, I begin lining a wall by sighting it quickly for bows. My partner nails braces where the deviations are worst. Then we brace off the rest of the wall where needed. If the wall is pretty straight, we scatter braces every 10 ft. or so, and do them in order.

The top of the line brace is attached to the wall first. Face-nail it to a stud just under the top plate with two 16d nails, making sure that it doesn't run beyond the exterior plane of the wall. If a header is in the way, face-nail a vertical 2x4 cleat to the header with three or four 16d nails, and then edge-nail the brace to the cleat. Line-brace connections are shown in the center and bottom drawing panels on the next page. Trimming an eyeballed 45° angle on the top of the brace will allow it to hug the wall and give you better nailing.

Setting the braces also requires good communication. While you pay attention to the string, your partner either pulls or pushes the brace according to your directions, and then anchors the bottom end when you say that it's looking good. You can usually find an inside wall to tie the bottom of the brace to. This won't affect plumbing the inside wall (which will be done later) as long as you nail the line brace to the bottom of a stud down where it butts the bottom plate. Use at least two 16d

Scissor levers for lining stubborn walls

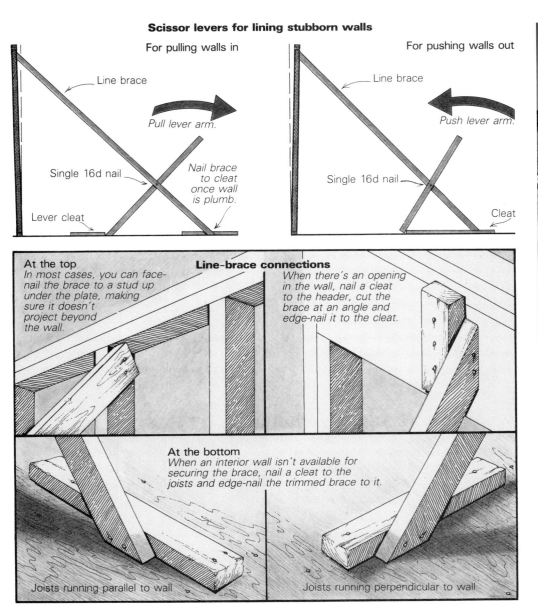

For pulling walls in

Line brace

Pull lever arm.

Single 16d nail

Nail brace to cleat once wall is plumb.

Lever cleat

For pushing walls out

Line brace

Push lever arm.

Single 16d nail

Cleat

Line-brace connections

At the top
In most cases, you can face-nail the brace to a stud up under the plate, making sure it doesn't project beyond the wall.

When there's an opening in the wall, nail a cleat to the header, cut the brace at an angle and edge-nail it to the cleat.

At the bottom
When an interior wall isn't available for securing the brace, nail a cleat to the joists and edge-nail the trimmed brace to it.

Joists running parallel to wall

Joists running perpendicular to wall

Persuading stubborn walls. Using a 2x scrap (a 3-footer is ideal) for a kicker under a line brace will bring a wall in at the top. Straightening walls quickly requires the carpenter setting the line braces to make adjustments in or out a little bit at a time with the constant direction of his partner who is eyeballing the wall.

Lining. For absolute accuracy, there is no substitute for a dryline. But if you've got a good eye and an experienced partner you can get straight walls in a hurry (facing page). Either way, use the top exterior corner of the double top plate to gauge the straightness of the wall.

nails (or duplex nails) and drive them home. If an inside wall is not available, nail or pin a block to the floor, no closer to the outside wall than the wall's height. This will leave the brace at an angle of about 45°. Be sure to find a joist under the floor, since a floor cleat that is nailed only to the plywood will pull out. Trim the bottom of the brace in the same way you did the top, so that you can drive two good 16d nails through it into the cleat.

Most walls are relatively easy to move in or out, but every house has a notable exception or two. Walls full of headers can be a real pain to line because they are so rigid. Bringing the top of a wall back in is usually more difficult than pushing it out because you have to work exclusively on the inside of the wall. One effective technique is to use a *kicker* (photo above). Toenail the line brace flat (3½-in. dimension up) to the header and nail a block above it so it won't pull out under tension. Then toenail the bottom of the brace into a joist, and cut a 3-ft. 2x scrap. Toenail one end of it to the floor and wedge the top under the line brace so you have to beat on the kicker to get it perpendicular to the line brace. This bows the brace out, bringing the wall back in

(although occasionally it pulls the toenails loose on the line brace). When the wall lines up, end-nail the brace to the kicker.

You can also use a parallel interior wall to help pull in an exterior wall, or use a scissor lever that nails temporarily to the line brace. Scissor levers can be used to push or pull, depending on where the blocks are nailed (top drawing panel, above).

Line the exterior walls one at a time until you come full circle. Try not to block off entries and doorways with braces, but don't skimp either. I've learned to double the normal bracing on exterior walls that will carry rafters for a vaulted or cathedral ceiling. Also, if this ceiling includes a 6x ridge beam or a purlin 18 ft. or longer, I like to place the beam on the walls just after I've plumbed them. This way you don't have to pick your way through a forest of braces with that kind of load.

Interior walls—The last step is to plumb up the inside walls. Start at one end of the house and work your way through. You'll develop a sense of order as you go, and soon find that the remaining walls are already nearly plumb, and that the frame is beginning to become rig-

id and to act as a unit. Quite a few of the interior walls will have let-ins; if they don't, use the 8-ft. and 10-ft. 2x4s as diagonal braces on the inside of rooms. Where there are long hallways, I usually cut 2x4 spreaders the exact width of the hall measured at the bottom, and nail them between the top plates every 6 ft. to 8 ft. The spreaders will keep the entire hall a uniform width, and when the wall on one side is plumbed, the wall opposite will be too.

Once all the interior walls have been plumbed, go back to any long walls to check for straightness, and then throw in one more line brace for good measure. This is also the time to make a last check on exterior walls to make sure the interior-wall plumbing you did hasn't thrown a hump in the works. Then go around the entire job, shaking walls to make sure there's no movement. This kind of precaution means that when you're rolling joists or cutting rafters you won't have to measure the same span or run every few feet along a wall for fear of a bow, or worry that the walls have acquired a lean over the weekend. □

Don Dunkley is a framing contractor and carpenter in Cool, Calif.

Truss Frame Construction
A simple building method especially suited to the owner-builder

by Mark White

Standard frame construction is complex and can be baffling to the first-time owner-builder. When the purposes and natures of foundations, sills, sill plates, floor joists, partition walls, studs, cripples, headers, ceiling joists, top plates, rafters and sheathing are taken one at a time they can be understood. But novices have a hard time handling the complexity once they are staring at all the pieces on their sites.

I have been teaching building on the college level for the past six years, trying to find a method that would reduce house construction to simple elements. Having tried balloon framing, platform framing, tilt-up walls, post and beam, and variations of them all, I now think I've found an answer: truss frame construction.

The truss frame is not new. Contractors, and individuals all over the world, have fiddled with the idea for years. It began to attract more attention in the United States when the Department of Agriculture's Forest Products Laboratory in

Madison, Wis., erected an experimental building that combined floor, walls and roof in single truss sections. I was immediately taken with the simplicity of the concept. It looked like a system that would enable the owner-builder to come up with a sound, useful structure on the first attempt, given some basic training and guidelines.

To test this building concept, I sketched up a set of plans and ordered the appropriate lumber from one of our local sawmills. The lumber was ready in January. In February, I began work on the foundation, and I set the sill timbers in early March. The building, which I planned to build alone to see how well the system would work, was to be used as a rental unit. It would have 12-in. floors, 10-in. walls and a 12-in. roof, all stuffed with a nominal 12 in. of insulation, for an insulation value of R-45.

Our climate is quite mild, as Kodiak, an Alaskan island in the north Pacific, is warmed by the Japanese current. Winter temperatures rarely

drop to 20°F, and average between 30°F and 40°F most of the time. Still, the winters are long and our primary heating fuel is oil, which is delivered by tanker from the lower forty-eight. The price of oil hasn't gone down in years, which leads us to think hard about proper insulation.

Foundations—Concrete costs a lot up here (about $165 a cubic yard), so many foundations are either creosote posts or treated wood. I opted for posts, because this is the fastest method and disturbs the soil the least.

The frost line here is 6 in., and bedrock is usually between 18 in. and 36 in. below the surface of the soil. I dug into glacial till—a mixture of hard clay and shale gravel a few inches above the actual bedrock.

The posts are Douglas fir, pressure-treated with creosote and about 12 in. in diameter. The ends that go into the holes are cut off squarely, then covered with another coat or two of creo-

The author built his first truss frame house alone to test the simplicity of the technique. At left, he winches the completed trusses onto sills set atop a post foundation. Center, a framed and sheathed partial truss is tilted into place as an end wall. It will be toenailed and temporarily braced. Right, most of the trusses are

sote to protect the center where the pressure treatment has failed to reach. I then nail a few pieces of heavy asphalt shingles (smooth side in) over that end to keep water from wicking up through the center of the post. When the post is in the hole, an eventual burden of 6,000 lb. to 10,000 lb. of house forces the asphalt into the wood fibers of the end grain and pretty much seals the pores. It takes 20 or 30 years to rot out the untreated center of a Douglas fir post, but I've seen it happen. The houses we build should last at least 200 years—for this reason I'm interested in having the posts last that long as well.

We carefully dig holes by hand into the 6-in. layer of glacial till and pour about a gallon of clean dry sand into the bottom of the hole instead of using a concrete pad. The sand is easier to work with if the posts need to be shifted to line up properly, and we have had little evidence of settling. We wrap the part of each post that is going to be in the ground with a few layers of 6-mil polyethylene to reduce the leaching of poisonous creosote into the groundwater. The plastic wrapping would deteriorate rapidly in sunlight, but it lasts a long time underground.

After the plastic goes on, the post is dumped into its hole, rotated a half turn in the sand and then propped into correct alignment with a few wedges jammed into the hole on one side or another. Once all the posts are in position, they are aligned with a transit, and everything is tied together with rough-cut 2x6s fastened with hot-dipped, galvanized, 20d nails. Extensive cross-bracing is installed before any weight is placed on the piles. The bracing is extremely important, because the soil is so shallow that it lends

little racking resistance to the system. Once the bracing is in, the holes can be filled and tamped around each post. Sand makes the best fill, but we usually wind up using the dirt that came out of the hole.

We usually space posts 6 ft. o.c., forming "strings" of them to support two 8x12 sill timbers along the length of the building. Near each end, we reduce the spacing to 3 ft. or 4 ft. o.c. to support the greater weight of the end walls and the extra load transmitted to them by the roof overhangs. We space the parallel strings of posts between 14 ft. and 20 ft. apart, depending on the carrying capacity of the floor joists the sills are supporting. Post foundations on sand and bedrock work well as long as the quality and spacing of the individual members is kept within reason, and the cross-bracing is adequate. We tend to be conservative, planning shorter spacing than the maximum indicated by charts and tables. A foundation is not worth skimping on. Besides, our cost only runs between $200 and $300 per structure—dirt cheap compared to the cost and labor associated with concrete.

The first house—Teaching duties and a building project in a remote village kept me from further work on the truss house until May. Then I cut out and assembled a single truss on the sill timbers. I used it to pull master patterns, from which I then traced the necessary shapes on 10-in. and 12-in. rough-cut spruce planks. I used a portable circular saw and a small chainsaw to cut out the pieces for the rest of the trusses.

The Forest Products Lab's original truss frame design called for the use of standard 2x4 mate-

rial for all chords and webs and results in many more pieces in each truss. I stayed with 1½-in. by 10-in. and 1½-in. by 12-in. material in the interest of simplicity. My trial structure was to be 20 ft. wide by 24 ft. long, with outward-sloping walls and generous porches all around. The outward-sloping walls were an experiment aimed at providing more visual interior room for a given area of floor space. They did provide the room, but for a few dollars more the side walls near the floor could have been kicked out a bit under the same roof and I would have had even more room. In a word, the experiment was successful, but I wouldn't repeat it. A frame spacing of 24 in. called for 11 full trusses, two partials for the end walls, and a total of four roof trusses to support the porch overhang at the ends.

A partial truss is a truss that has gussets on one side and studs on the other side. Trusses are used on the end walls only to define the shape and outline of those walls and to hold the studs in that configuration before the walls are tilted up into position.

Roof trusses are made up of rafters and collar ties, without vertical members. If the overhang at the gable end is less than 3 ft., it is possible to use roof trusses supported only by a sturdy 2x12 fascia board nailed to the other rafters along the eaves. If the overhang is greater than 3 ft., some arrangement of conventional headers and cripples is necessary. This often makes even the cross tie unnecessary.

Construction of the trusses went according to plan, but including a floor joist as a part of each unit turned out to cause more trouble than it was worth. It meant that there was no floor to work

erected, toenailed to the sills, and stabilized by the plywood nailed along their sides. In this first house. floor joists were a part of each truss. This resulted in great strength and stability, but made construction a bit awkward, since there could be no platform to work on until all the trusses were in place.

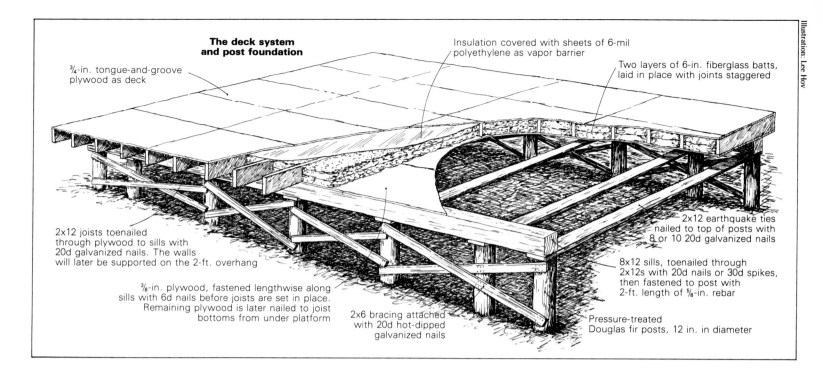

Illustration: Lee Hov

The deck system and post foundation

¾-in. tongue-and-groove plywood as deck

Insulation covered with sheets of 6-mil polyethylene as vapor barrier

Two layers of 6-in. fiberglass batts, laid in place with joints staggered

2x12 joists toenailed through plywood to sills with 20d galvanized nails. The walls will later be supported on the 2-ft. overhang

⅜-in. plywood, fastened lengthwise along sills with 6d nails before joists are set in place. Remaining plywood is later nailed to joist bottoms from under platform

2x6 bracing attached with 20d hot-dipped galvanized nails

2x12 earthquake ties nailed to top of posts with 8 or 10 20d galvanized nails

8x12 sills, toenailed through 2x12s with 20d nails or 30d spikes, then fastened to post with 2-ft. length of ⅝-in. rebar

Pressure-treated Douglas fir posts, 12 in. in diameter

on, so moving materials, assembling them and erecting them was awkward work.

The building did go together with less effort than one built with either the standing stick or the tilt-up wall method. And the truss assembly with its plywood gussets made an extremely strong and rigid structure. If anything is wind and earthquake-proof, it is this building. I was heartened by the progress and by the rigidity of the building, but felt the system could be simplified even further.

The completed first house. Sloping side walls increase the volume with limited floor space.

An improved design—The following fall, I designed a 20-ft. by 26-ft. house with straight walls 6¼ ft. high. This time the floor would be built and insulated as a separate unit (drawing, above), with the trusses and end walls erected on top of it. This is much easier, and almost as strong.

It took an inexperienced, se n-person crew (my class) six hours to build and insulate the floor. First, we laid our joists (in this case 20-ft. long 2x12s) over two sill timbers, spacing them 24 in. on center. Then we nailed ⅜-in. CDX ply-

wood to the joists from underneath and dropped in two layers of 6-in. fiberglass insulation, making sure to stagger their joints. On top of this went a 6-mil vapor barrier, then either a plank or a plywood floor. Dropping insulation in from above is an easy 10-minute task. If we had had to install it from below, it would have become a frustrating, eye-irritating chore.

Each truss designed for this building consists of five main pieces: two wall studs, two rafters and one collar tie. The class cut and assembled the required twelve full trusses, two end walls and three roof trusses in exactly 14 hours of hard labor. To eliminate inaccuracies, each truss was assembled on top of the master truss. All of the ⅝-in. plywood gussets were nailed with great quantities of #6 galvanized nails.

I have used gussets of ½-in. plywood, but they sometimes break when they're fastened to only one side and the truss is being flipped over to have the rest nailed onto the other. Using ⅝-in. plywood has eliminated the problem. I would recommend ¾-in. plywood for gussets on larger trusses and in two-story houses. Once gussets are nailed onto both sides of a truss, breakage is extremely unlikely, even with ½-in. plywood.

Truss members are very thick to provide room for lots of insulation. The trusses are thus massively overbuilt, so a builder need not worry much about a structural failure. A neophyte designing a truss pattern need only include a substantial enough collar tie to keep the rafters from spreading out near their bottoms. The collar ties should equal about half the span of the rafters. We usually use 2x12 ties, which can safely span 16 to 18 ft. On longer spans, it's safer and cheaper to use smaller dimensioned lumber, with a center tie fastened between the center of the collar tie and the peak of the roof. The center tie adds strength and keeps the ceiling from sagging and the long collar ties from warping.

We scheduled raising the frame for a Saturday so the students could work all day. Moving the sections and clearing ice from the floor's platform took about two hours. Putting up the end walls and the trusses themselves took 17 minutes. Passers-by were amazed at the speed with which the building took shape.

Initially, the bottoms of the studs were toe-nailed to the floor, with each stud positioned over a floor joist. The first course of plywood sheathing was nailed to the floor perimeter and to the studs as the trusses were raised. This course secures the studs to the deck and keeps the trusses erect. More plywood sheathing on the walls and 1x6 decking on the roof locks everything together.

Alignment of the walls and roof was perfect. This turned out to be the straightest and squarest building I've ever seen assembled—thanks to a carefully leveled floor and to the uniformity of the trusses.

Insulation—The class met for four hours one final Saturday to get the building weathered in. To insulate the roof, we strung nylon twine under the rafters and stapled it in place to hold the fiberglass. In about 15 minutes, two layers of 6-in. insulating batts were laid in from above. We then decked over the roof with 1x6 spruce, trimmed and covered it with a layer of 55-lb. roofer's base felt. This layer will eventually be covered with asphalt shingles, but it will protect the building until good weather returns.

We have built a number of buildings with floor, wall and roof sections insulated to R-45. They are effective, allowing us to heat the average home with 15 to 20 gallons of fuel a month during the winter. A conventionally insulated house of equivalent size may gobble up to 350 gallons of fuel a month during a similar heating season.

In most of these buildings, I used Owens Corning 6-in. fiberglass in the roof. It seems to come 5½ in. thick, so two layers gave me 11 in. of insulation and a 1-in. ventilation space. I used only friction-fit batts or rolls, no foil face. In the floors and walls, I used Johns-Manville 6-in. fiberglass, two layers of which usually measure out to 13 in. (Owens Corning and Johns-Manville claim the same R-value, and they cost the same.) Recently I have switched to 12-in. wall cavities to make better use of the Johns-Manville insulation.

Plumbing, electrical and heating—The structures my classes and I have built appear deceptively simple in shape, form and function. They are not. Their cross-sectional designs have been carefully worked out to provide maximum floor space and volume with minimum exposed surface area. There are a number of deviations from typical construction practice that could be imposed upon any building method, but seldom are. The very nature of the truss frame structure, along with our floor-building technique, simplifies and encourages their use.

First, the floor is an almost totally sealed unit. Its vapor barrier is penetrated only by a carefully installed waste pipe and a water supply line. We use 6-in. or 8-in. interior plumbing walls, and try to back the kitchen up to the bathroom, utility room and wash room to get all the plumbing into one area.

Under the house, I make no attempt to insulate the 3-in. waste drain that flows to the sep-

Trusses for the second house did not include floor joists. A deck was built first so that work could progress easily, then—in just 17 minutes—trusses were tilted up, toenailed and temporarily braced, before being firmly connected to the deck by the first course of horizontal plywood sheathing. These workers are stringing twine beneath the rafters to hold 12-in. batts of fiberglass insulation.

tic system. It is merely angled properly and enters the ground quickly. None of the many that have been installed this way in our location has ever frozen. I usually wrap the 1-in. PVC water supply line with a short length of thermostatically controlled heat tape extending just below the frost line, and then insulate it. A neon indicator light on the upper end of the tape tells you whether it's working or not.

In Alaska, we install this 1-in. pipe inside a 2-in. pipe, which in turn should be insulated to beneath the frost line with heavy, black neoprene foam made for the purpose.

Electrical power enters the house by way of a single piece of 1½-in. conduit passing from a meter base on an outside wall to a single service panel on the wall inside the vapor barrier. All circuits emanate from this service panel, passing throughout the building in a single channel cut into the interior face of the outer wall studs. (There are two such channels cut into the faces of the studs in a two-story dwelling.) We use either Romex cable or conduit, ½ in. or ¾ in. in diameter. All cables and conduits are inside the vapor barrier, so there are no leaks through the membrane. Switches and electrical outlets don't let cold air get into the room.

The heavy insulation and complete vapor barrier eliminate drafts, convective air currents, and excessive heat losses, so there is little need in our houses for complex heat distribution systems. For heat sources we have tried wood-stoves but they typically put out too much heat. We have settled on either a standard oil-fired, hot-air furnace, or an oil hot-water heater with a short loop of pipe run from the water tank to heat the air. In fact, we have a real problem finding an appropriately sized heat source. Right now the smallest furnace available is in the neighborhood of 85,000 Btu. What we really need is one that kicks out 12,000 Btu or less.

Potential uses—The frame truss system lends itself to simple buildings with repetitive sections, and I've used it to build structures with conventional shapes. But different applications and shapes are possible, because the basic truss is highly adaptable. I once built a strong and very lightweight building out of 1½-in. by 1½-in. stock and plywood glued and nailed in place. If you live in an area where labor is considerably cheaper than material you might try a truss composed of 2x2s or 2x3s and light plywood gussets glued and stapled in place to form an intricate webwork. This would eliminate the problem of direct heat transfer through solid joists, studs and rafters. It would, however, introduce the new problems of insulating between the webwork and of sealing off passages to fire and rodents. In our area rough-cut lumber is available at a reasonable price in lengths of up to 24 ft. so trusses of solid lumber are more cost effective than lighter, more intricate designs.

The use of a well-ventilated, heavy truss system instead of concrete in an underground structure merits consideration. The strength of a properly designed truss makes it a good choice in the high load conditions found beneath the earth. Another area of application would be in passive solar designs where the shape would be in one of the many asymmetric configurations, with the tall open side facing south.

The beauty of the truss system is that only one of the many frames that go into the building needs to be laid out with great care. Once that first unit is formed it's an easy and repetitive task to construct the rest of the units, using the first as a pattern to get the rest right. Raising the frame is then a simple matter involving a minimum of fiddling and measurement. □

———————

Mark White teaches at the University of Alaska at Kodiak.

Making a Structural Model

Fine-tuning your design and working out construction details is a lot less expensive at ¹⁄₂₄ scale

by Mark Feirer

If you've ever passed a construction site and paused to watch a framing crew, you may have marveled at the precise choreography of their work. A glance at the blueprints, and these carpenters are back to fast, precise cutting and nailing as if genetically encoded with the right instructions. Not all of us have such a sense of how a building's many structural parts fit together, and this explains the value of building a model before you're facing full-scale construction.

The framer's source of information is a set of blueprints, but even the best-drawn blueprints can't match the comprehensive impression a scale model conveys. A novice builder can be intimidated by a relatively simple house in the same way that an experienced contractor can be befuddled by an unusual design. Both can benefit from building a scale model. Models don't eliminate the need for working drawings, but they do increase the builder's understanding of a structure. If you believe that practice makes perfect, consider this: How else can you build your house twice and benefit from your mistakes without having to live with them?

The design model—This is a sort of three-dimensional sketch. Its purpose is to let you briefly explore a design before you make final decisions that will be translated onto your working drawings. It often takes several models before you can make a decision, so these models are characteristically inexpensive and quick to construct. They're usually built from sheet materials like mat board or cardboard at a scale of ¼ in. = 1 ft. (1/48 size).

The design model replicates only the shell of the building and doesn't attempt to answer structural questions. Nevertheless, because it's a scale replica of the real thing, it lets you test for shading and sun penetration quite accurately. Solar designers can place a model in a device called a heliodon, which simulates the path the sun takes at different latitudes and times of the year (Sun Light makes an elegant one). A design model can also be used to test the livability of a floor plan by adding interior partitions to it. With the roof removed, you can "walk" through the house.

Structural models—While design models show the skin of a house, structural models show a building's skeleton, and are therefore more useful from a builder's point of view. Constructing this kind of model enables you to trouble-shoot the structural design of a house, eliminating or revising framing techniques that are difficult or wasteful. As any builder knows, just because you can draw it doesn't mean you should build it that way. Complex or experimental structures can be examined at the modelmaking stage for flaws that might otherwise make site work a nightmare, or drive the cost beyond reason.

Engineers can test a structural model by subjecting it to scaled-down stresses of gravity, wind shear and seismic activity. Models given this type of testing have to be specially built. However, if your craftsmanship is tolerable and your finished model collapses when you accidentally bump the table, you might want to re-examine your design.

Another benefit of structural modelmaking is that once the model is finished, it can help you make a materials estimate for the building. All you have to do is count the number of studs, joists, rafters and other structural members. You can also determine the most efficient layout for sheathing and subflooring.

Building a structural model gives you a manageable replica to study before and during full-scale construction. Facing page: the author positions a roof truss on a model built to 1/24 scale (½-in. = 1 ft.). White or yellow wood glue, a razor knife or two, and a fine-toothed hobbyist's saw are some of the tools you'll need. Model construction follows the same sequence as actual construction, although it's possible to skip the foundation and start with floor framing, as was done here. Floor sections were only partially sheathed so that the floor-framing details would remain visible.

Scale and materials—The level of detail that you can achieve in a model depends largely on its scale. The detail can increase as the scale gets larger, but at some point the model can get unwieldy and expensive without really offering more advantages than you'd have at a smaller scale. For structural models that replicate an entire house, a scale of ½ in. = 1 ft. (1/24 size) is a good compromise: it's large enough to show joist cross-bridging but small enough for tabletop work. At this scale, individual pieces of the model are usually large enough to handle without tweezers. If you want more detail, 1/12 scale (1 in. = 1 ft.) is the logical next step up. You can buy modelmaking materials that are predimensioned to both these scales.

Modelmaking materials and tools are available at most hobby shops (sometimes in the guise of model-railroad or model-aircraft supplies) and some stores that supply architects and architecture students. Materials may be purchased by mail order as well, from Northeastern Scale Models, Inc. (P.O. Box 425, Methuen, Mass. 01844) and Eugene Toy and Hobby (32 E. 11th Ave., Eugene, Ore. 97401).

Spruce, basswood and balsa are the most popular modelmaking woods, and you can also find predimensioned plastic extrusions. Balsa is the lightest, weakest and easiest to work, and it comes in the widest variety of forms: sheets, slabs, scaled-down 2x lumber and blocks. Unless the work calls for extra strength or stiffness, most of the structural members in our models are balsa.

Model lumber is scaled to the nominal dimension of the full-size item; a 2x4 will scale to 2 in. by 4 in., not 1½ in. by 3½ in. The difference is barely noticeable at 1/24 size. Model woods are usually found in lengths of 2 ft. or 3 ft. (48 ft. and 72 ft. in 1/24 scale) so a considerable amount of cutting to length is required before you can begin gluing.

You can, of course, dimension and prepare your own materials from scrap stock, but I wouldn't recommend doing so unless you're long on time and short on funds. However, if you need an off-sized piece and you can find balsa sheets in the thickness you want, it's easy to rip individual framing members to width with a razor knife and straightedge. You could also use redwood, cedar, yellow poplar or fir, as long as the wood is dry.

Adhesives take the place of mechanical fasteners in modelmaking. White or yellow wood glues make a strong bond and get tacky quickly. It's usually only 10 seconds before a piece dabbed with a pinprick of glue will stay in position without being held, and after 10 minutes the bond is fairly secure. Sometimes balsa requires two coats of glue because it's so porous, particularly in end-grain joints.

Testor Corp. (620 Buckbee St., Rockford, Ill. 61101) makes an extremely strong and quick-drying adhesive called Cement for Wood Models (modelers sometimes call it Green Label). Though its unpleasant odor and relatively high cost ($.59 for a ⅝-oz. tube) count against it, wood glued with Testor's develops a strong bond in just 15 to 20 seconds. For an even faster bond, I've had success using the cyanoacrylate "super glues" that sell for about $3 per ½-oz. tube. They're expensive, but they bond almost instantly.

Tools—The razor knife is the portable circular saw of model building—you can't do without it. I use two: a lightweight knife with a scored blade that can be snapped off when it gets dull, baring a fresh edge; and a standard mat knife, like the one Stanley makes, for heavier cutting. A straightedge is the companion to the razor knife. I use a metal one 18 in. long and 1½ in. wide. The third essential tool is an architect's scale, so that you can dimension structural members for cutting. The standard triangular-section scale works fine.

A miter box and a fine-tooth modeler's saw aren't essential, but they are very useful. The miter box we use is an aluminum, hobby-shop model with slots for both 45° and 90° cuts. A pair of dividers or a compass with two metal tips is great for laying out dimensions and marking wood to be cut. Even thin pencil lines are usually too wide to use in marking small pieces, so the pinprick left by dividers is useful. A large hat pin comes in handy for this same kind of scribing and for applying glue.

In addition to these basic tools, most modelmakers end up making or improvising a number of their own. When you're working at such small scale, it's amazing what household or workshop items get pressed into service: straight pins for temporarily fastening wood to wood, paper clips (which can be bent to form fine clamps) and long-nosed tweezers. The X-acto Co. (4535 Van Dam St., Long Island City, N. Y. 11101) makes a variety of useful blades to fit their basic razor knife. Dremel Co. (4915 21st St., Racine, Wis. 53406) makes such exotic modelmaking tools as miniature power saws and lathes.

Estimating materials—If you already have a materials estimate for the full-scale house, just use the lumber portion of it as your shop-

From *Fine Homebuilding* magazine (August 1984) 22:50-53

Using a cut-off jig like the one shown above enables you to mass-produce studs or joists of uniform length. The jig has a thin plywood bottom that can be wedged in different positions in the metal miter box, and a glued-down stop for the stock to butt against.

The wall-assembly jig, top, consists of two fixed, parallel guideboards that are spaced the same distance apart as the full plate-to-plate height of the wall. To use the jig, top and bottom plates are aligned against the guides and then studs are positioned over perpendicular layout lines and glued at top and bottom to both plates. Once the glue is dry, completed wall sections can be stood up and glued in place or glued to scaled-down sheathing or siding, as shown above. A razor knife is your rip-saw at ¼₄ scale.

ping list for model materials—with one adjustment. Convert any listing of individual pieces into a total lineal-foot measurement for each dimension category. For example, a listing of 200 lineal feet of 2x4s is more helpful than a listing of 20 10-ft. 2x4s because model lumber usually isn't scaled to length. I add a waste factor of 5% to 8% to my materials list.

If you don't have a materials take-off for the plan yet, make a quick estimate for the model. You can figure liberally in this case. Carrying an inventory of ¼₄-scale lumber over to your next model won't create a cash-flow problem. Once the model is complete, it will help you make an accurate list of framing and sheathing materials for the real house.

Before you buy materials, you have to decide how closely you intend to simulate the full-scale construction process. If your plans call for a plate on one wall composed of two 16-ft. 2x4s, will you use that combination on the model or will you use a continuous plate? Using the continuous plate will save time, but there's a lot to be said for building to the plans, especially if this is your first house. The more you can learn from your model, the better. As you work, imagine yourself doing the same thing at the site. It's easy to forget how heavy a 24-ft. 6x10 beam really is when you're lifting its counterpart with tweezers.

The model base and foundation—You need a flat, rigid working platform to build on, preferably one that's easy to mark up so you can outline your foundation. Particleboard and plywood (with one A face) both work well. Use ¾-in. material to get the rigidity you'll need to take your scale creation on site for reference. I make the base 2 in. to 3 in. larger than the overall size of the model, and sometimes allow a few extra inches on one side to serve as a cutting area for materials. Lumberyard remainder bins are a fairly reliable, low-cost source of base stock.

For the sake of authenticity, it's best to construct your model in the same sequence you'd follow in building a real house: foundation, floor system, first-floor walls, second floor and so on. Taking the measurements from the plans, draw an outline of the foundation on the base, representing the outside face of the foundation wall. Include the location of any interior support walls, piers or other structural members.

If I'm planning to build on a level site, I usually consider the surface of the model base to represent the top of the footing, and build foundation walls up from there. For hillside sites, it's possible to create a scaled-down slope by layering mat board (or something similar) on a flat base to correspond with actual site contours. If you don't foresee any problems with the foundation because it's straightforward, you can skip this step by using the top surface of the base as the top of the foundation. This way you'll begin by laying out the mudsill.

Double-check your foundation or perimeter-wall layout against your plans—square corners and accurate dimensions are just as im-

portant now as they are at full scale. If you elect to build a foundation for your model, particleboard or plywood are good materials to work with. At ¼₄ scale, thicknesses of ¼ in., ⅜ in. and ½ in. correspond to full-size wall thicknesses of 6 in., 8 in. and 12 in., respectively. Cut the foundation walls to proper height and length with a saw and then glue them in place on your layout lines. Butt-joined corners are fine, but mitered corners give the model a cleaner look.

Floor systems—Whether your plans call for 2x6 T&G decking over girders or ¾-in. plywood over 2x joists, you should be able to duplicate this construction at ¼₄ scale. To represent decking over girders, you can either lay down individual pieces or use a thin (¹⁄₁₆-in.) sheet of balsa or model-aircraft plywood, which is far stronger. Sometimes called wingskin, model-aircraft plywood is available in thicknesses ranging from ¹⁄₆₄ in. to ¹⁄₁₆ in. It cuts easily with several passes of a sharp razor knife, but to make short work of cutting individual sheets to consistent size I use a paper cutter.

Joisted floor systems take a bit more work. Once you've cut the joists to length, mark their locations on the plates or mudsills with dividers set to the center-to-center spacing of the joists. Prick a small hole with the point of the dividers at each joist location, and don't forget to indicate doubled joists, headed-off joists and other specials that are shown on your blueprints. Now dab the joist with a pinhead of glue and set it on its mark. Fight the temptation to use more glue than you need; a little goes a long way. Once you've got several joist runs glued down, eyeball them to make sure that they are running true.

To sheathe a floor or wall, you can use either sheet balsa or model-aircraft plywood. You might not want to sheathe your model completely at the floor, walls or roof, since this will cover many of the structural parts that you'll want to look at later.

Two jigs—To build our models, I use two jigs to speed construction and ensure accuracy: a cut-off jig for cutting studs to length, and a wall-assembly jig for putting all the parts together quickly and accurately.

The fastest way to cut modeling lumber to length is with a cut-off jig. The one I use is essentially a small, adjustable table that fits inside my model-maker's miter box, and has a stop block fastened permanently to one end (photo top left). The table is made from ¹⁄₁₆-in. aircraft plywood; it also protects the sawblade by preventing it from touching the metal base of the miter box.

To use the jig, mark a single stick of material to the length you want and position it with one end against the stop block and one edge against the side of the miter box. Slide the jig forward or back until the stick's cut-off mark lines up exactly with the blade-guide slot in the box. Then shim the plywood table firmly against the side of the miter box to hold its position—a business card or two should be

Rafters as well as roof trusses are modeled flat on a template or assembly jig and then glued in place atop the top plates, as shown at right. A time-saving alternative is to model the trusses from solid pieces of plywood or cardboard.

enough if the table has been sized correctly. Now cut your test piece and check its length. If it's right, you can load the box with five or six random lengths of material and gang-cut them to size. Just make sure that the stock is aligned straight and butted against the stop block as you cut. Scribing registration marks on both jig and miter box allows you to return the table to a given position without having to re-measure. Variations of this jig allow you to make repetitive cuts at different angles.

The wall-assembly jig is basically a platform about 8 in. long and an inch or so wider than the finished wall height. Two parallel wood guides are glued to the top and bottom of the platform; the distance between them should equal the wall height, from bottom plate to top plate. Perpendicular layout lines drawn on the jig between the guides correspond to the on-center spacing of the studs (photo facing page, center). Horizontal lines can be added to locate blocking or headers.

Using the jig is not much different from building a full-scale stud wall on a completed subfloor. Cut a bottom plate and a top plate to the length of your wall and align them on edge against the parallel guides. Now glue in your studs, dabbing a pinhead of glue on both ends and aligning each one on its proper layout line. The fit of the studs in the jig is important, so I cut them after it is built. I do this because it's easier to adjust the studs to the jig than vice versa, and getting the height of the wall correct is the most critical factor. The fit of the studs should be snug enough that they stay in place (before they're glued) when the jig is turned upside down, but if they bow (bowing is particularly easy with balsa), they are too tight. Keep a piece of 150-grit sandpaper nearby to make minor adjustments. Walls of most common lengths will fit within the 8-in. length of the jig. But you can also build longer walls simply by sliding completed portions through the jig.

Installing walls—A wall can usually be installed on the subfloor as soon as it leaves the assembly jig, though I like to let one wall assembly dry while I'm building the next one. When the second one is finished I install the first, and so on through all the first-floor walls and partitions.

The first step in installation is to mark the location of interior-partition walls because this is easiest to do before perimeter walls go up. Exterior walls can be installed using the outer edge of the subfloor as a guide; I usually begin with a pair of adjoining exterior walls.

To attach a wall to the subfloor, simply hold the wall in position after dabbing the sole plate (bottom plate) with glue. Walls are usually straight and square as they come from the assembly jig, but you can use a straightedge to take out the bow in a plate. Once two exterior

walls are up I install any partitions that attach to them. This construction sequence does two things: The partitions brace the exterior walls and help to hold them in place should I accidentally bump one. And I'm also spared the tricky maneuver of reaching between completed exterior walls to install interior partitions, which can demand the patience and finesse of a surgeon.

Other installations—The second and third floors are built in the same way, using whatever scaled-down components the plans call for. If walls of a different height are required, you have to modify your wall-assembly jig, or build a new one.

Builders often use floor trusses to frame the second story, because they allow long clear spans. You can model these in two ways: as a solid member or as a truss. The quickest and easiest method is to model the truss as a solid, using a 1/24 scale 2x member with the same depth as the truss. To model it as a truss with all its webbing, I use a modified wall jig.

You can prepare rafters with an angled cut-off jig similar to the straight cut-off jig mentioned earlier. A fairly accurate bird's mouth can be cut in each rafter with a razor knife,

though you can't lay it out with a framing square. The best approach is to make a template rafter by setting a single piece of rafter stock in position against the ridge and the top plate and marking it for plumb and seat cuts. Cut and adjust the fit with a razor knife until you get a close-fitting template, then use it to pattern the other rafters.

Roof trusses (photo above) can be fairly complex, and you might not want to spend the extra time modeling them exactly if you're planning to use prefab units. If this is the case, you can use one of four options: 1) Cut a solid piece of aircraft plywood or balsa to represent each truss; 2) build just the outer chords of each truss, leaving the webbing out; 3) ignore the roof framing structure altogether and build a tent-type roof out of mat board with cleats glued to the underside to anchor this sheathing to the top plate; 4) or build one or two trusses, webbing and all, as illustrations of how they would look, and build the rest of the roof using one of the other techniques. Any of these approaches will save time, particularly if at this point you're anxious to see the overall shape of the house. □

Mark Feirer lives in Eugene, Ore.

The Engineered Nail

Ten steps to secure connections

by Edward Allen

The nailed connections in an ordinary house frame conform to a chart found in most building codes. Most designers and builders seldom give them a second thought. But many building designs contain connections that aren't of the ordinary sort—columns to major beams, connections in trusses, splices in roof ties and unusual intersections in floor framing. And many times a nailed connection must be designed for some non-standard project such as a footbridge, a large wind-buffeted sign, a piece of playground equipment or a temporary exhibit structure.

Too, a builder often designs scaffolding, concrete formwork and other one-of-a-kind temporary structures in which nailed connections must be counted on to carry substantial loads without failing. In any of these special situations, it's reassuring to know that nailed connections can be engineered with the same degree of precision as connections made with bolts, lag screws and split rings.

All things considered, nails are the most versatile fastening devices available for wood construction. They are inexpensive and lightning fast to install; they are always available in the sizes that we use most; and they are less obtrusive than bolts and lag screws. As a bonus, nailed connections are a lot less complex to engineer than bolted or screwed connections.

General considerations—The design of nailed connections is governed under all of the major building codes by the *National Design Specification for Wood Construction*, which I will refer to in this article as the NDS. It's the bible of engineering practice for wood construction, and is published by the National Forest Products Association (1250 Connecticut Ave. N. W., Washington, D. C. 20036). You may find it worthwhile to become familiar with this book if you design wood buildings, because it covers calculations for beams, joists, columns, tension members, glue-laminated wood, and connec-

Structural connections in wood, like those shown for these roof trusses, can be engineered with precision and confidence. By knowing how much force the connection must transmit and being able to determine what the holding power of each nail is, you can design nailed connections for everything from carrying beams to cupolas. The simplified look-up tables in this article were compiled from data published in the *National Design Specification for Wood Construction.*

tions using bolts, wood screws, lag screws and split rings, as well as nails.

But to use the NDS effectively you need to have had a course or two in structural engineering. For this article, I have summarized its provisions on designing connections using common nails, and have tried to put its guidelines into a form that's easy to use.

There are some things to keep in mind when you set out to design a nailed connection. Com-

mon nails, box nails, spikes and threaded, hardened-steel masonry spikes and nails can all be used in structural connections under the NDS. But finish nails can't be used as structural fasteners.

Box nails are weaker than common nails, so if you want to design a connection using box nails you will have to refer to the NDS, as you will for common spikes, which have somewhat higher values than those given here for common nails. Vinyl-coated sinker nails don't yet appear in the NDS, so we have no firm basis for designing with them. For this article, threaded hardened-steel spikes and nails have the same load-carrying capacities as common nails.

A word of caution here. Never use drywall screws as structural fasteners. Drywall screws are good for hanging gypsum board and for non-load-bearing connections in cabinetwork. But they are much weaker than nails or common screws, and they are very brittle. Even a properly installed drywall screw will fail without warning at a distressingly low loading.

It is best to design nailed connections so the nails are loaded laterally rather than in withdrawal, as shown in figure 1 on the facing page. A nail that is loaded in withdrawal from end grain is prohibited by the NDS, and a nail loaded in withdrawal from side grain, while acceptable, is discouraged. Withdrawal loadings can almost always be avoided by using overlapping connections, scabs or gussets of plywood or metal.

Designing a connection—To design a nailed connection, proceed step-by-step as follows:

Step 1. Sketch the connection, and indicate with arrows the loads or forces the connection must transmit. Make the joint as simple as you can, and design it so that it does not use nails in withdrawal. As a general rule, avoid using nails to "hang" a beam or joist onto the side of a column without support from below. A beam held up by its own nails alone is liable to fail in

From *Fine Homebuilding* magazine (June 1987) 40:68-72

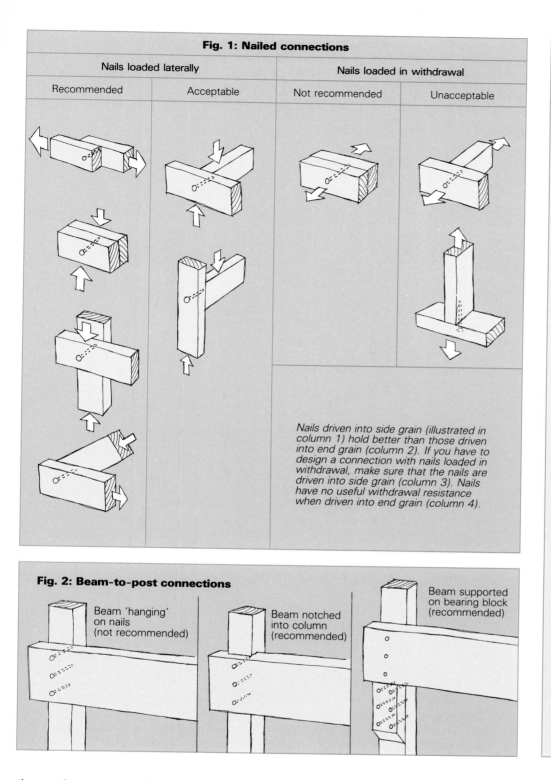

Fig. 1: Nailed connections

Nails loaded laterally		Nails loaded in withdrawal	
Recommended	Acceptable	Not recommended	Unacceptable

Nails driven into side grain (illustrated in column 1) hold better than those driven into end grain (column 2). If you have to design a connection with nails loaded in withdrawal, make sure that the nails are driven into side grain (column 3). Nails have no useful withdrawal resistance when driven into end grain (column 4).

Fig. 2: Beam-to-post connections

Beam 'hanging' on nails (not recommended)

Beam notched into column (recommended)

Beam supported on bearing block (recommended)

Species group

Commercial species are rated for hardness, ranging from group I (ash, maple, oak, etc.) to group IV (various softwoods).

Ash, commercial white: I
Aspen: IV
Balsam fir: IV
Beech: I
Birch, sweet and yellow: I
Black cottonwood: IV
California redwood: III
California redwood, open grain: IV
Coast Sitka spruce: IV
Coast species: IV
Cottonwood, eastern: IV
Douglas fir / larch: II
Douglas fir, south: III
Eastern hemlock: III
Eastern hemlock / tamarack: III
Eastern softwoods: III
Eastern spruce: IV
Eastern white pine: IV
Eastern woods: IV
Engelmann spruce / alpine fir: IV
Hem / fir: III
Hickory and pecan: I
Idaho white pine: IV
Lodgepole pine: III
Maple, black and sugar: I
Mountain hemlock: III
Mountain hemlock / hem-fir: III
Northern aspen: III
Northern pine: III
Northern species : IV
Northern white cedar: IV
Oak, red and white: I
Ponderosa pine: III
Ponderosa pine / sugar pine: III
Red pine: III
Sitka spruce: III
Southern cypress: III
Southern pine: II
Spruce-pine-fir (SPF): III
Sweetgum and tupelo: II
Virginia pine / pond pine: II
West Coast woods (mixed species): IV
Western cedars: IV
Western hemlock: III
Western white pine: IV
White woods (western woods): IV
Yellow poplar: III

shear at the connection. (For an explanation of beam shear, see "Building Basics," pp. 8-12.)

There are a couple of ways to avoid hanging a beam, as shown in figure 2. One is to set the beam into a notch in the column so that it can transfer its load by direct bearing, and use the nails only to hold the beam in the notch. The other is to use nails to attach a bearing block to the side of the column, then to support the beam on this bearing block and keep it in place with nails. This kind of support minimizes the shear problem and allows you to use as many nails as necessary to carry the load.

Step 2. Determine how much load the connection must carry. If the connection is in a roof structure, add up all the loads that must pass through the joint you are designing, including the live loads required by your building code and the weight of the structure itself.

If you're designing a floor structure, the live load is usually figured as 40 lb. per sq. ft. under most building codes for a residential structure, but is generally more than this for commercial, industrial and agricultural buildings. For scaffolding, add up the largest total load of workers, materials, equipment and scaffolding planks. Then figure out what portion of this weight each connection must support.

Step 3. Determine the species group of the wood for which you are designing. For simplicity, the NDS uses density as a criterion to divide all species of wood into four groups. The denser the wood, the more load each nail can carry. All the species in Group I are hardwoods. Group II includes the strongest softwoods, and Group III consists of the majority of the softwood species used for framing lumber. Group IV includes still other species that are often used in construction for framing.

The chart above is an alphabetical list of common woods used in building structures, indexed by their species as shown on lumber-grade stamps, and assigned to a group (I, II, III, IV) according to density. Be sure you select the right species group for your design, because nail load-bearing capacities are heavily dependent on wood density. If you are in doubt about what wood species you are designing for, assume it belongs to Group IV, just to be safe.

Step 4. Select the size of nail you would like to use in the connection. In most cases you will

Photo and drawings: Edward Allen

DESIGN VALUES IN POUNDS:
Lateral resistance of common nails in side grain

Depth *	Species group I				Species group II				Species group III				Species group IV			
	Nail size				Nail size				Nail size				Nail size			
	8d	10d, 12d	16d	20d	8d	10d, 12d	16d	20d	8d	10d, 12d	16d	20d	8d	10d, 12d	16d	20d
½ in.	37	39	0	0	27	0	0	0	0	0	0	0	0	0	0	0
⅝ in.	46	49	51	0	34	36	38	0	23	0	0	0	17	0	0	0
¾ in.	56	59	62	67	41	43	45	49	28	30	31	0	21	22	0	0
⅞ in.	65	69	72	78	47	50	53	58	33	35	36	40	24	26	27	0
1 in.	74	78	82	90	54	58	61	66	37	40	42	46	28	29	31	34
1⅛ in.	83	88	92	101	61	65	68	74	42	45	47	51	31	33	35	38
1¼ in.	93	98	103	112	68	72	76	82	47	50	52	57	35	37	38	42
1⅜ in.	97	108	113	123	74	79	83	91	52	55	57	63	38	40	42	46
1½ in.	97	116	123	134	78	87	91	99	56	60	63	68	42	44	46	51
1⅝ in.	97	116	133	146	78	94	99	107	61	65	68	74	45	48	50	55
1¾ in.	97	116	133	157	78	94	106	115	64	70	73	80	49	52	54	59
1⅞ in.	97	116	133	168	78	94	108	124	64	75	78	86	51	54	58	63
2 in.	97	116	133	172	78	94	108	132	64	77	83	91	51	59	62	68
2⅛ in.	97	116	133	172	78	94	108	139	64	77	88	97	51	61	66	72
2¼ in.	97	116	133	172	78	94	108	139	64	77	88	103	51	61	69	76
2⅜ in.	97	116	133	172	78	94	108	139	64	77	88	108	51	61	70	80
2½ in.	97	116	133	172	78	94	108	139	64	77	88	114	51	61	70	85
2⅝ in.		116	133	172		94	108	139		77	88	114		61	70	89
Maximum drill size	⁷⁄₆₄-in.	⅛-in.	⁹⁄₆₄-in.	¹¹⁄₆₄-in.	³⁄₃₂-in.	⁷⁄₆₄-in.	⅛-in.	⁹⁄₆₄-in.	³⁄₃₂-in.	⁷⁄₆₄-in.	⅛-in.	⁹⁄₆₄-in.	³⁄₃₂-in.	⁷⁄₆₄-in.	⅛-in.	⁹⁄₆₄-in.

* Depth of penetration of nail into member receiving the point.

want to use the largest size that is reasonable (the chart at right gives lengths and diameters for common nails).

For example, if you are face-nailing nominal 2-in. pieces of lumber to one another, 10d nails make the most sense because their length is just equal to the overall thickness of the connection. For end nailing, 16d or 20d nails are better because you get greater penetration.

Step 5. Determine the penetration of the nail into the second wood member. By using the sketch you have made, you can see exactly how much of the length of the nail you have selected is used up in going through the first piece of wood into which it is driven, and how much of the nail this leaves for holding power in the second piece.

Step 6. Look up the design value for the nail in the chart above. Start by finding the correct species group for the wood you are using at the top of the table. Next, within that species group, find the size of nail you are using. Then read down that column in the table until you come to the line that corresponds to the penetration depth you determined in Step 5.

The number at this location in the table is the design value in pounds for the lateral resistance of a common nail that has been driven into side grain. (These design values are based on laboratory tests of nailed connections, and include a factor of safety). Write down the value you find there and proceed to the next step.

Step 7. Now you'll need to modify the design value you have just found in accordance with NDS requirements. These requirements can be summarized as follows: If a nail is driven into end grain, its holding power is diminished, and so you'll have to multiply the design value by

Dimensions of common steel wire nails		
Nail size	Length	Wire dia.
Eight penny (8d)	2½ in.	0.131 in.
Ten penny (10d)	3 in.	0.148 in.
Twelve penny (12d)	3¼ in.	0.148 in.
Sixteen penny (16d)	3½ in.	0.162 in.
Twenty penny (20d)	4 in.	0.192 in.

0.67. If it's a toenail, which is also weaker than a face nail, multiply the design value by 0.83.

The tabulated design values are for connections in which primary loading is typical of residential or commercial floor loads—the weight of people and furniture. If the major portion of the load on the nail comes from roof loads (snow or rain), multiply the design value by 1.15. For a connection made to resist wind or earthquake loadings, multiply the design value by 1.33. On the other hand, if the entire calculated load is permanently applied to the nail, multiply the design value by 0.9. These duration-of-load factors reflect the fact that wood can support short-term loads better than long-term loads. For floors, use a factor of 1.0, for roofs, 1.15, and for diagonal bracing, 1.33. An example of a permanent load, with a factor of 0.9, is support for a hot tub or water tank sitting on a deck.

When using unseasoned or partially seasoned wood, or if the wood in the connection will be damp or wet in service, multiply the design value by 0.75. If you are using fire-retardant, treated wood, multiply the design value by 0.9 (a connection in preservative-treated wood is considered to be as strong as one in untreated wood).

If the joint is made with metal gusset plates through which the nails are driven, multiply the

design value by 1.25. If the nail will be in double shear, through side pieces each at least ⅜ in. thick, is not over 12d in size, and is clinched at least three nail diameters, as shown in figure 3, at the top of the facing page, multiply the design value by 2. Notice that other than this provision of NDS, there is no increase in design value allowed for clinched nails.

These modifications are cumulative. This means that if, for example, you are designing a rafter connection between unseasoned wood members using toenails, you should multiply the tabulated design value for the nail by 1.15 for roof loading, then by 0.83 for toe nailing, then by 0.75 for unseasoned wood. The net effect of these computations in this example is that you will multiply the design value by 0.72.

Step 8. To get the required number of nails in the connection, divide the total load on the connection by the modified design value per nail and round upward to the nearest whole number. The strength of a nailed connection is equal to the sum of the strengths of all the nails in the connection. (This may seem patently obvious, but research has shown that it is often not true in bolted connections, for example.)

Step 9. Go back to your original sketch and lay out the locations of the nails in the connection. If only two or three nails are required, this is easy. Frequently, however, your calculations will show the need for a dozen nails or more in the same connection, and you'll need a scale drawing or full-size layout to figure out how to place them.

There is no hard-and-fast rule about nail spacings. The NDS specifies only that nails be spaced so as to avoid excessive splitting of the wood. As a rule of thumb, many architects and engineers try to space nails no closer together than the length of penetration at which they achieve their maximum design value. Consider using a 10d nail in a Group III species. The design-value chart shows that it reaches its maximum design value at a depth of 2 in., which can then be taken as the minimum spacing between nails. For the distance from the outermost nails to the sides of the piece, half this value is acceptable. The distance from the last nail to the end of the piece should be equal to the required spacing.

However, certain connections often call for more nails than this rule of thumb would allow. When this happens, prepare a sample connection with scraps of the same wood that will be used in the finished structure to see if splitting is a problem. If it is, try staggering the nails in such a way that they do not line up closely along the same grain line, as shown in figure 4. Drilling pilot holes for the nails further reduces the possibility of splitting.

Step 10. If you decide to drill lead holes, you'll have to select the right-size bit. The holes must have a diameter large enough to ease nailing and reduce wood displacement, but not so large that holding power is lessened. To find the correct size of bit to use (as recommended by the NDS) look at the bottom of the design-value chart. In sketching the connection, be sure to dimension the nail spacings you have arrived at, and indicate the diameter of drill bit that should be used for the lead holes. And remember that while the NDS allows soap or grease to be used on screws, this doesn't apply to nails in structural situations, because nails depend on friction for their holding power. Use no grease.

Example 1: A beam-to-post connection— Having gone through all the steps, let's design a connection to support the end of a double 2x12 floor beam that is sandwiched around a 6x6 post, and assume that both beam members and the post are KD Douglas fir. The first thing to do is *sketch the connection*. To avoid hanging the beams and the consequent shear problems, you can add nailed-on bearing blocks, as shown in the drawing at bottom right, which also prevents weakening the post by notching into it.

Next, you have to determine the load. Let's assume that calculations have already been done that show the total load on the beam to be 3,380 lb., uniformly distributed along its length. Half of this amount, or 1,690 lb., must be borne by the post at each end. Because the beam is double, each 2x12 brings half this amount, or 845 lb., to the post—so 845 lb. is the design load for each connection.

Third, look up the species group of the wood in the chart on p. 63, where you'll find that Douglas fir is a Group II wood. Now select the nail size. The post is thick enough to accept 20d nails readily, and if you drill pilot holes, there is little danger of splitting either the block or the post with such large nails.

Now you need to determine how deeply the nails will penetrate. A 20d nail is 4 in. long, and the bearing block is 1½ in. thick, leaving a penetration into the post of 2½ in. Next look up the design value for the nail in the design-value chart. Under Group II species, a 20d nail pene-

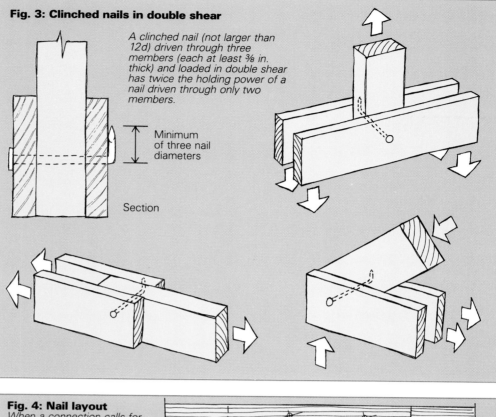

Fig. 3: Clinched nails in double shear

A clinched nail (not larger than 12d) driven through three members (each at least ⅜ in. thick) and loaded in double shear has twice the holding power of a nail driven through only two members.

Minimum of three nail diameters

Section

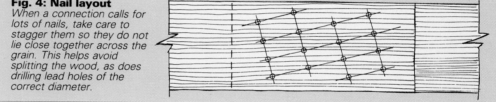

Fig. 4: Nail layout
When a connection calls for lots of nails, take care to stagger them so they do not lie close together across the grain. This helps avoid splitting the wood, as does drilling lead holes of the correct diameter.

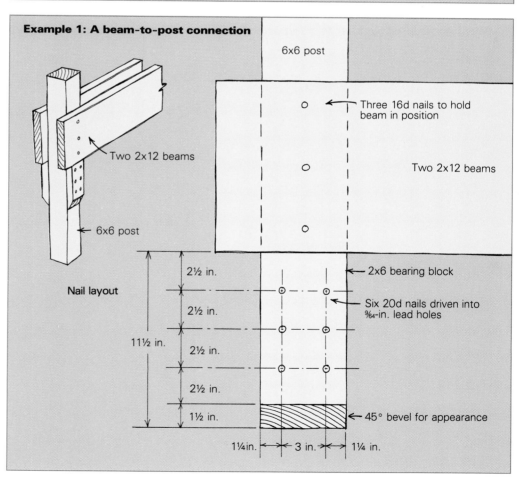

Example 1: A beam-to-post connection

6x6 post

Three 16d nails to hold beam in position

Two 2x12 beams

Two 2x12 beams

6x6 post

Nail layout

2½ in.

2½ in.

2½ in.

2½ in.

1½ in.

11½ in.

2x6 bearing block

Six 20d nails driven into ⁹⁄₆₄-in. lead holes

45° bevel for appearance

1¼ in. — 3 in. — 1¼ in.

trating 2½ in. will support 139 lb. Design values usually have to be modified, as discussed above, but in this example, none of the modification factors apply; so we will proceed with a load-bearing capacity per nail of 139 lb.

Now you can figure out how many nails the connection needs by dividing the 845-lb. load by the 139 lb. that each nail will carry. You end up needing 6.08 nails. You could be conservative and use 7 nails, but 6 is close enough in this case (less than 2% overstress), especially when you consider that you'll also be attaching the beam to the post with three nails. These will give you a wide margin of safety.

Now you can lay out the detail of the connection, as shown in the drawing at the bottom of the previous page. Since 20d nails reach full bearing value at a penetration of 2⅛ in. in Group II woods, a spacing between nails of 2⅛ in. is about right, with an edge spacing of half this amount. The bearing block is 5½ in. wide and 11½ in. long. In laying out the detail of the connection to scale, you can of course increase the spacing if it improves the way the thing looks. The nails through the sides of the 2x12 beams are intended mainly to keep the beam in place, and not to transmit the load from the beam to the column.

The last thing you need to figure is the diameter of the bit you need to drill the lead holes. In the design-value chart we find that a ⁹⁄₆₄-in. drill bit is suited for the lead holes. A larger bit would result in insufficient friction between the wood and the nail.

Example 2: A roof-truss connection—The problem this time is to design an end connection in a roof truss to be made of KD spruce-pine-fir (a common type of framing lumber cut from stands of trees of mixed species in the Pacific Northwest). The loads on the truss come primarily from snow. Assume that you've already consulted the building codes and done the math to figure the maximum design loads for the top and bottom chord and the weight the stud below must bear. The amounts and directions of the forces are indicated in the drawing below left. The wood belongs to Group III. We select a 10d nail because its length will just reach through the thickness of the joint. The penetration into the second member is 1½ in.

From the design-value chart, you determine that the design value per nail is 60 lb. This is modified for snow loading by multiplying it by 1.15 to yield a value of 69 lb. per nail. Now you need to look at the sketch of the connection (drawing, below left) to figure out which of the three forces to design for. The 375-lb. force is carried by direct wood-to-wood bearing of the truss on the top of the wall, and needs no nails except to keep the truss in place. The 745-lb. diagonal force of the rafter is counteracted in part by the 375-lb. vertical bearing force, and in part by the 643-lb. horizontal force in the bottom chord of the truss. The only force not supported at least partly in direct wood-to-wood bearing is the 643-lb. horizontal force, so this is what we must design the nailed connection to resist.

Dividing 643 lb. by 69 lb. per nail gives a requirement of 9.33 nails, which we round up to 10. The rule-of-thumb nail spacing in this situation is 2 in., the penetration at which 10d nails reach full capacity in Group III woods. But trying to lay out ten nails in this connection with this spacing is fruitless. And going to a larger nail won't help much either, because at 1½ in. of penetration even a 20d nail holds only 8 lb. more than a 10d nail.

One thing you can try is to increase the size of the bottom chord to a 2x6 to increase the area that will receive the nails. Such an increase in member size is often necessary when designing wood connections. Alternatively, if Group II woods such as Douglas fir and Southern pine are available, you can try the design again in these species to see if the reduced number of nails and reduced rule-of-thumb spacings result in a workable design.

But let's go back to the original design and see if we can squeeze the nail spacings to make it work. One of the virtues of the lower-density woods of Groups III and IV is that many of them, like spruce-pine-fir, are very soft and don't split easily. In the proposed layout (drawing, left), the nail locations are staggered in such a way that they minimize splitting forces, and you'll reduce splitting further if you predrill the nail holes with a ⁷⁄₆₄-in. bit as indicated by the design-value chart. Now mock up the assembly with scraps of your framing lumber and see if any splitting happens. If the sample shows signs of splitting, you will have to resort to an alternate design. If the layout works, make a template of the spacings so that you can locate the holes uniformly on the real connections. □

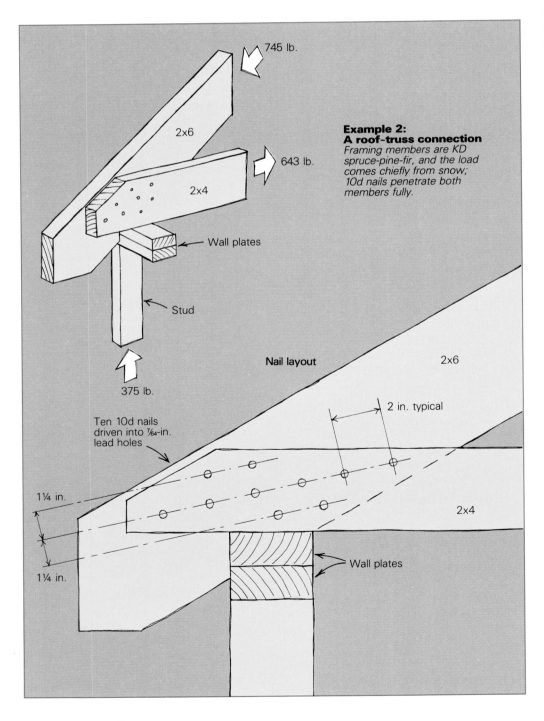

Example 2: A roof-truss connection
Framing members are KD spruce-pine-fir, and the load comes chiefly from snow; 10d nails penetrate both members fully.

745 lb.

2x6

643 lb.

2x4

Wall plates

Stud

375 lb.

Ten 10d nails driven into ⁷⁄₆₄-in. lead holes

Nail layout

2x6

2 in. typical

2x4

1¼ in.

1¼ in.

Wall plates

Edward Allen is an architect in South Natick, Mass. He is the author of Fundamentals of Building Construction *(John Wiley & Sons, 605 Third Ave., New York, N. Y. 10158).*

Toenailing

by Albert Treadwell

One of the most frequently ignored devices in all of carpentry is the simple nail. Everyone takes it for granted. Everyone thinks that everybody else instinctively knows what nail to use where, and how to drive it properly. Even on the few occasions when the experts do decide to write about nailing, they rarely say much about toenailing. For example, a letter in *Fine Homebuilding* magazine from engineer/builder Don McNeice treated load engineering and design considerations for fastening joists to beams and headers, but he did not address the role of toenails, a vital element of any nail-fastened structure.

McNeice correctly points out that spikes into the end grain of a framing member have no withdrawal resistance, and it's pretty clear that a lag screw won't grab in end grain either. Friction can be an important part of a joint's strength, above and beyond the sum of the shear strength of the individual nails mandated by the design load. Properly installed toenails will pinch a joint tight, and their shear strength will help keep it tight (drawing, right).

The importance of accurate cutting to length and a good square end cannot be overemphasized, and this alone is enough to warrant using a radial-arm saw on site. All too often in today's (and yesterday's) construction, you find crooked or short rafters or joists with contact reduced to a line or a single point. Some even stand open completely, with nail shanks visible. Nails in such a joint are subject to stresses considerably more complex than simple shear, and the members are free to twist and move.

You could drill to lessen the chance of splitting, but in the case of NcNeice's face-nailed spikes this would also decrease the already minimal compression-friction grab the joist end offers the nail. On the other hand, you can drill toenails for the full diameter of the nail, because the head is pulling the joint together. When you're toenailing, you don't need to drill into the second piece, because the face grain usually takes nails well. It is possible by pre-drilling the board to

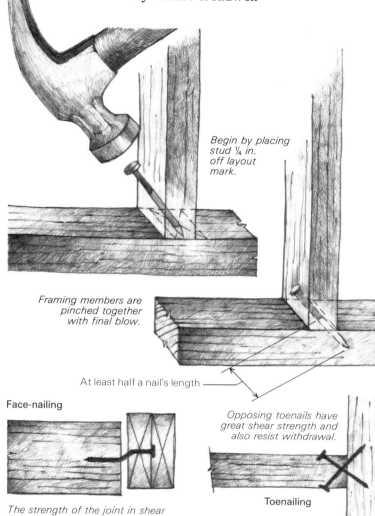

Begin by placing stud ¼ in. off layout mark.

Framing members are pinched together with final blow.

At least half a nail's length

Face-nailing

The strength of the joint in shear relies solely on the strength of the nails or spikes. There is little resistance to withdrawal.

Opposing toenails have great shear strength and also resist withdrawal.

Toenailing

Frances Boynton

place toenails anywhere you like without splitting the wood. You can drive as many as six toenails in a 2x4, eight in a 2x6, ten in a 2x8, and so forth.

Now an 8d or 10d toenail doesn't offer the shear strength of a face-nailed 20d or 30d spike, but ten 8d toenails and one or two spikes in a 2x8 will make a stronger and more durable joint than peppering the same joist with four or five spikes driven into the end grain. The strength of such a joint is simply the sum of the shear strengths of the spikes. It relies on strapping or the nailed-down floor for its integrity. The toenailed joint adds friction, developed in the joint by the pulling action of the opposing toenails, itself a result of horizontal shear in the small fasteners. Also, with toenails on all sides, force from any direction will make the joint tighter and stronger.

How to do it—Many experienced framers don't use toenailing either because they haven't thoroughly understood its practical value or because they never learned efficient technique. Poorly done, toenailing will result in split wood, misaligned work, and joints standing open, to say nothing of smashed fingers and hammers thrown into the woods.

Toenailing is not difficult and can even assist in the final alignment and positioning of the piece. To begin, place the stud or joist end in position, but off its layout line by ⅛ in. to ¼ in. Start a toenail on one side while bracing the other with your hand or foot. You may start the nail perpendicular and swing it to the correct angle, but it must be placed so that more than half of its length will enter the second piece. The nail must enter the piece to which you are joining

the stud or joist before the stud or joist is pushed into its final position. The final hammer blows will move the piece onto the line, and this is what creates the pinching action that forces the pieces tightly together. The distance you start from the layout line is critical. With some practice, you'll be able to determine this based on the size of both members and the length of the nails. Too little distance and you will either generate no pinching action or overshoot the mark, causing the piece to stand off the joint when you drive it back. Too much distance and you won't make it to the mark, and you may even split the piece by driving the head of the nail too far.

The sequence of subsequent nails is determined by the alignment corrections that are necessary. Remember that when the piece is held by one nail, it can pivot, but it won't move if force is aimed directly at that nail. So if the piece is a little crooked, you have to position the next toenail to bring it around. If you are right on, aim the next nail almost directly at the first from the other side. Continue with the remaining nails in like fashion.

Another subtlety concerns the final hammer blow to each nail; if it strikes both the nail and the wood, the piece will move. This is useful when correction is necessary. If the final blow of the hammer strikes only the nail, the piece will not move. A hammer without a crowned face and beveled edge (like my old Stanley #41) is useful here. So is a heavy nail-set (I use a ¼-in. mechanic's punch).

In a sensitive spot, it pays to drill a pilot hole. Some would say this is a waste of time, but a split destroys any strength the joint might have had. You either have to replace the split piece or leave a structurally inadequate member in place. I like to use a cordless drill, because being tethered to an outlet can be irritating and dangerous. The small holes require only a little power, so a single charge should last all day. For smaller jobs a hand drill or push drill is good. □

Contractor Albert Treadwell lives in Sandy Hook, Conn.

Metal Connectors for Wood Framing

Galvanized steel folded into sturdy, time-saving devices

by Bruce R. Berg

A swallowtail scarf joint is an elegant way to make a tension tie between adjoining beams, and a housed dovetail will anchor a floor joist to a girder for the life of a structure. But not many construction budgets have an allowance for the extensive and meticulous cutting and fitting that it takes to achieve these time-honored joints. These days, most structural connections in wood-frame buildings are made with steel connectors because they are affordable and easy to install. Also, their structural values have been carefully tested and documented. Consequently, steel connectors are widely accepted by codes and building officials.

The companies that make metal connectors (see the sidebar on p. 73 for sources of supply) offer their products in a remarkable number of configurations. Their catalogs include not only illustrations of the connectors, but also tables that list specifications such as the dimensions of the lumber and the appropriate connectors, their design loads, and the number and size of nails it takes to achieve that rating.

Joist hangers are probably the most common type of metal connector on a construction site, but if you need them you can get connectors to anchor a scissors truss to a bearing wall, adjustable post bases or metal clips that allow you to install outdoor decking without visible nails. This article takes a look at the principal types of steel connectors. Within these categories there are many variations that you can use to solve specific construction problems.

Concrete-to-wood connectors—If you have ever struggled to lift a framed wall onto a protruding row of anchor bolts, you are familiar with a potential source of frustration. Despite everyone's best intentions, the holes in the sill plate sometimes don't line up with the bolts, and the plate has to be redrilled. Or a stud lands on an anchor bolt, requiring a nasty-looking notch in the bottom of the stud.

One alternative to anchor bolts that circumvents these problems is the MAS galvanized steel anchor from Simpson (drawing A, facing page). It resembles a Y with a ladle-like cup on the bottom leg that gets embedded in concrete. The branches of the Y are wrapped around the mudsill or up the side of a stud and secured with nails. Prior to the pour, these anchors can be positioned by tacking them to the formwork. And because they emerge from the concrete at the edge of the footing, you don't have to hand-trowel around a bunch of anchor bolts.

Another sheet-metal anchor from Simpson (called the MA) can be attached to the mudsill before you pour your foundations (drawing B). The anchor's pointy, arrowhead shape allows it to slip easily into the screeded wet concrete.

Seismic anchors (sometimes called hold-downs or tie-downs) are frequently specified by architects and engineers when part of a structure needs lateral bracing and there is only a narrow wall section in which to provide it. The narrow wall is stiffened with plywood for shear

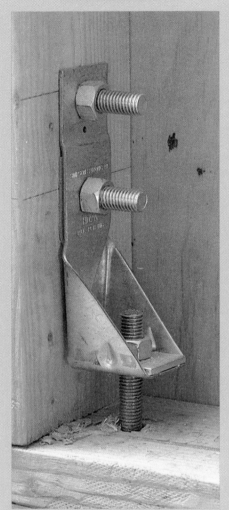

Hold-downs are tension ties between the foundation and the framing. The distance of the bolts from the ends of the studs must be at least seven times the diameter of the bolt. Often engineers will specify greater distances.

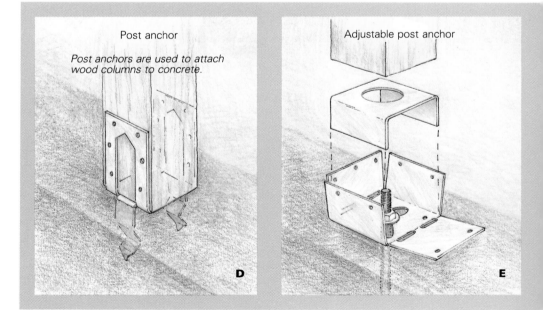

Post anchor

Post anchors are used to attach wood columns to concrete.

D

Adjustable post anchor

E

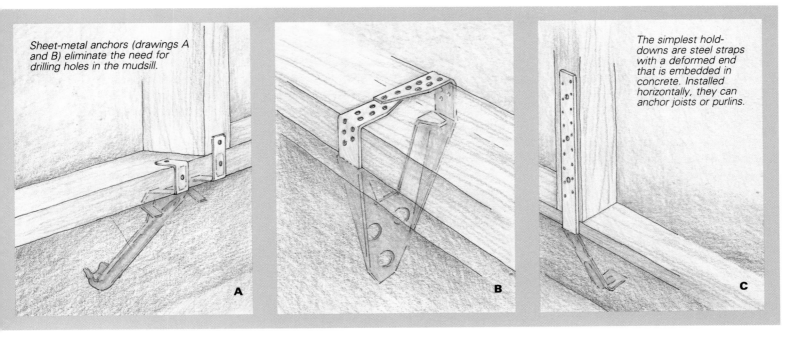

Sheet-metal anchors (drawings A and B) eliminate the need for drilling holes in the mudsill.

A

B

The simplest hold-downs are steel straps with a deformed end that is embedded in concrete. Installed horizontally, they can anchor joists or purlins.

C

strength, but when a horizontal load is applied to the top of a stiff panel, it wants to lift away from one of its corners. A hold-down provides resistance to this uplift.

Hold-downs come in two basic varieties. The first is a deformed strap that is set into the wet concrete at the location of a post or stud (drawing C). It is then nailed or bolted to the post, sometimes with as many as 24 nails.

The other variety uses a foundation bolt that rises through the plate and through the bottom of a gussetted, welded heavy angle to which it is fastened (photo facing page). The angle is then bolted to the stud or post at the perimeter of the plywood panel. Because of the seismic activity here on the West Coast, builders frequently use pairs of these hold-downs linked by a threaded rod to create a tension tie between two floors.

Very similar to the deformed-strap hold-down are purlin and joist anchors. They are embedded horizontally into the concrete or masonry wall,

aligned with the top of each framing level. When nailed off to the joists or purlins, they allow the horizontal diaphragm to work together with the wall structure, and prevent the walls from leaning outward.

A peek under many backyard decks will reveal row upon row of concrete pier blocks supporting a forest of 4x4 posts. To distribute their loads evenly, the blocks are typically set into a bed of wet concrete. A chunk of pressure-treated pine or redwood is attached to the top of the block for toenailed connections. Pier blocks are easy to work with, but they have their limitations. The wood blocks sometimes detach from the concrete, and they are so small that they often split when you drive nails into them. Also, they provide little resistance to lateral or uplift loads, and can't be replaced if required. A far better way to anchor posts, such as those used to support decks, is the post or column base.

Post anchors are made in several configura-

tions—each for a particular application. One kind is designed to be placed into wet concrete (drawing D, facing page), and its positioning must be precise. Another type uses an anchor bolt to secure it to concrete, and it has a slotted adjustment plate (drawing E) that allows you to tinker with its alignment in the event of a slightly misplaced bolt. You can also use this type of post base on a cured concrete slab or footing by tying it down with an expansion bolt, concrete nails or powder-actuated fasteners.

Some post anchors have a standoff plate that elevates the post about 1 in. above the concrete, which prolongs the life of the post in moist areas. Under really wet conditions, you can use an elevated post anchor (drawing F) to get the wood several inches above the concrete.

Cleveland Steel makes a dressy post base out of cast aluminum that lends a sturdy appearance to a column, and elevates the wood above the concrete (drawing G). The base is secured with

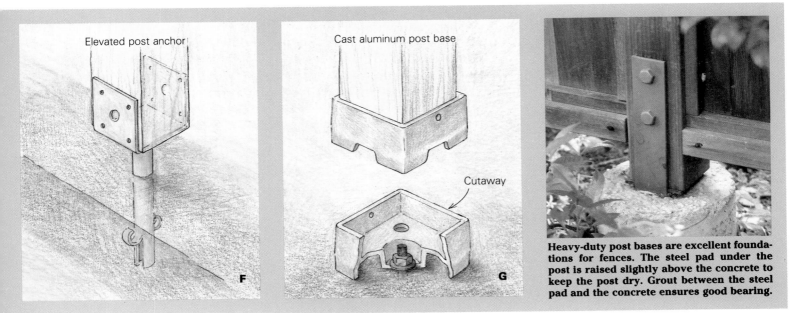

Elevated post anchor

F

Cast aluminum post base

Cutaway

G

Heavy-duty post bases are excellent foundations for fences. The steel pad under the post is raised slightly above the concrete to keep the post dry. Grout between the steel pad and the concrete ensures good bearing.

a single anchor bolt, and weep holes allow an escape route for rainwater.

Since most post bases allow little room for adjustment once they've been installed, you've got to be scrupulously accurate as you embed them in the wet concrete. Use a string line or a transit to align rows of post bases as they are set in the concrete. Lumber-crayon marks on the formwork can ensure the accuracy of your spacing in the other direction.

Sometimes builders use post bases to support beams, as on a low deck. In this situation you can first prepare the forms, then place the beam in its final position on top of falsework with the post bases attached to the beam. Pour your concrete, take out the falsework and your beam is ready to carry its load without the aid of a wooden column.

Post bases also come in heavy-duty versions. These are made of thicker steel than the standard bases, and have longer straps that extend up the sides of the post. One good application of the heavy-duty post base is to anchor fence

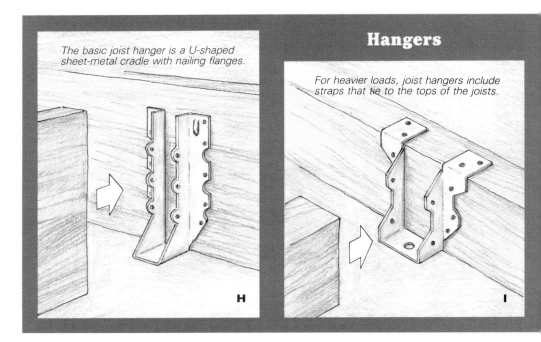

The basic joist hanger is a U-shaped sheet-metal cradle with nailing flanges.

Hangers

For heavier loads, joist hangers include straps that tie to the tops of the joists.

H

I

Straps and bracing

Steel straps are useful tension ties throughout wood-frame buildings. Here a steel strap is used to resist the uplift load on a beam.

posts (photo previous page). Excavate your post hole and use a short section of Sonotube to bring the level of the concrete about 6 in. above grade. As you fill the hole with concrete, insert a #4 rebar in the center of the hole before you embed the heavy-duty post base. Bolted to this kind of a base, your fence posts won't rot off at grade in ten to twenty years.

Hangers—The basic hanger is a galvanized strip of 14-ga. to 18-ga. steel folded into a U-shape (drawing H). Hangers are made in various sizes to accommodate typical framing lumber. At the bottom of the U, the metal widens to form a seat for the joist or purlin. Properly installed, the member rests snugly against the seat of the hanger. Flanges on the legs of the U turn outward, and nails driven through holes in the flanges secure the hanger to the beam; other holes allow fastening to the joist.

Representatives of the companies that make

steel connectors say that the most common mistake made in installing their products is inadequate nailing. Because the shear strength of the nails is the limiting factor in the strength of the connection, it's critical that you install the connectors with the nails specified by the manufacturer. Commonly available wire nails in the correct diameter are often longer than the thickness of the framing member, so you have to use "joist-hanger nails," which are simply short versions of common nails. Also, nails that are used in outdoor locations or driven into treated wood should be galvanized.

Another typical mistake is to cut the joists too short. The gap between the beam and the end of the joist should be no more than 3/16 in., and closer is better.

When you are installing joist hangers on a beam that is deeper than the joists, you have to snap a chalkline on the beam to align the bottoms of the hangers. Because joist material

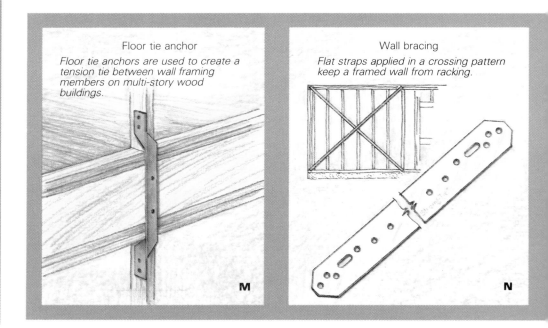

Floor tie anchor

Floor tie anchors are used to create a tension tie between wall framing members on multi-story wood buildings.

Wall bracing

Flat straps applied in a crossing pattern keep a framed wall from racking.

M

N

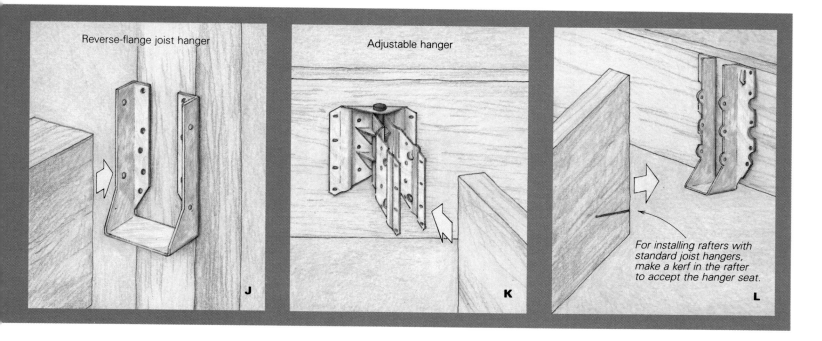

Reverse-flange joist hanger

Adjustable hanger

J

K

For installing rafters with standard joist hangers, make a kerf in the rafter to accept the hanger seat.

L

sometimes varies slightly in depth, position your hangers to accommodate the deepest joists in the lumber pile. Then add shims to the hanger seats to bring the shallower joists flush with the top of the beam.

To hang joists that carry heavier than normal loads, use hangers with a top flange (drawing I). When you nail down the subfloor, don't worry about the thickness of the metal draped over the joists—a few hammer blows will compress the wood fibers enough to make the floor lie flat.

On most hangers the flanges turn outward, but you can also get them with reverse flanges (drawing J). Reverse-flange hangers can be useful in tight spots, such as a window retrofit when you have to add a header and there isn't room for trimmers to carry its load, or when two perpendicular beams meet at one post.

Not all hangers are designed to carry members perpendicular to a beam. Most companies make 45° hangers as standard items, and some

manufacturers, such as Panel Clip and Cleveland Steel, are set up to make "specials" to fit unusual framing needs. They can make hangers that are sloped to carry rafters, skewed at angles other than 45°, or a combination of the two. Simpson makes a hinged hanger that will skew to 60° and slope up to 30° (drawing K).

Another way to handle a rafter with a conventional hanger is to cut a kerf in the end of the rafter that is the same depth as the hanger seat and perpendicular to the plumb cut (drawing L). The hanger seat tucks into the kerf.

Straps and bracing—Sometimes called tension ties, steel straps are applied to wood-frame construction in numerous ways. For example, if the plumber cuts through the top plate in a wall, a steel strap can restore the structural integrity of the plate without adding much bulk to the framing. Other common uses of the steel strap are to tie opposing rafters together across a

ridge, and to resist the uplift of a cantilevered beam (photo facing page).

One of the simplest yet most useful connectors I have used is the twist strap. A typical one is made of 16-ga., 1¼-in. wide galvanized steel bent so that the faces of the two ends are 90° to each other. They are often used in pairs on framing members that cross one another to prevent the wood from twisting. I've also used them as hangers to suspend old ceiling joists from new beams during remodel work.

Another version of the twist strap is the floor tie anchor (drawing M). A 90° degree twist at each end of this strap allows it to be attached to the framing on multi-story buildings to make a tension tie between floors. It does the same thing as a pair of hold-downs, for less money.

Metal wall bracing (drawing N) is another form of strap tie that can speed things up on a job site. While not nearly as strong as a well-nailed plywood diaphragm, it is at least as strong as 1x4 let-in bracing, and you don't have to cut notches to install it. To prevent racking, the flat variety needs to be applied in pairs to form an X or a V.

Whenever I'm remodeling older homes that have diagonal blocking, I enhance the shear strength of the wall by adding a flat wall brace to the line of blocks. I nail it securely to each block and each stud. If I'm working on an interior wall, I use a power plane to cut a shallow groove for the brace across the studs so it won't telegraph through the finished wall.

Another type of steel wall bracing comes in a T or L section (drawing O). It is let into a kerf cut across the studs in a straight line, providing the framed wall with both compression and tension bracing in a single strip.

Metal bridging (drawing P) is so much faster to install than wood bridging that it is an ideal example of how steel connectors have added to the efficiency of wood-frame construction. Prongs on the ends of the bridging eliminate the need for nails. You drive one end into a joist about 1 in. from its top edge, then the other end

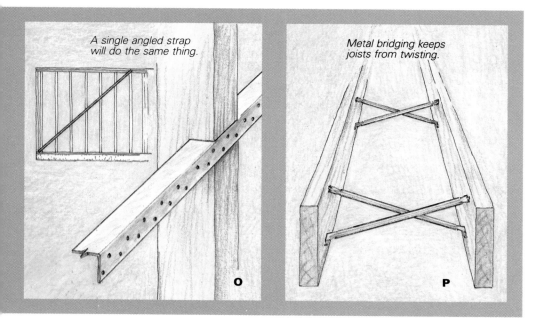

A single angled strap will do the same thing.

O

Metal bridging keeps joists from twisting.

P

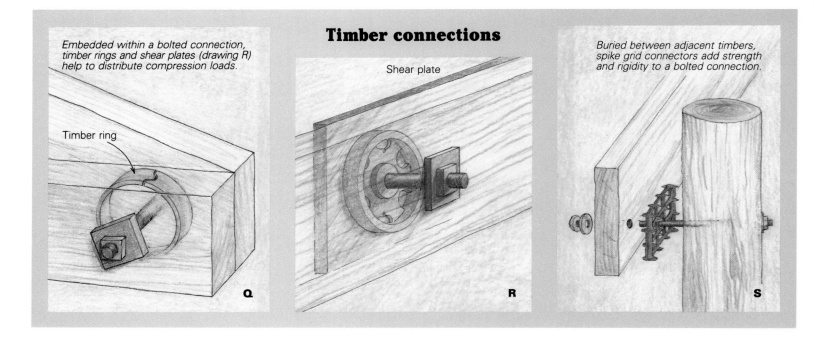

Timber connections

Embedded within a bolted connection, timber rings and shear plates (drawing R) help to distribute compression loads.

Timber ring

Shear plate

Q

R

Buried between adjacent timbers, spike grid connectors add strength and rigidity to a bolted connection.

S

into the adjacent joist 1 in. from its bottom edge. Because they require no nails, they don't develop nail squeaks. But don't let the pairs of bridging touch one another or they will make noise. Keep them 1 in. apart, and remember that on spans over 16 ft., you need to use two sets of bridging to conform to most building codes.

Timber connections—Timbers are typically bolted together, and because they are usually supporting substantial loads, a lot of pressure is concentrated on the bolted connections. Timber rings, also called split rings, and shear plates are two metal connectors that are used in concert with bolts to spread out compression loads and shear forces, reducing the potential for crushed wood fibers around the bolts.

Timber rings (drawing Q) are steel rings that ride in matching grooves cut into adjacent timbers. A grooving tool that resembles a hole saw is used to cut the grooves for the rings, and at

the same time it bores a hole for the bolt that runs through the center of the rings. When installed, the rings are hidden from view, captured by the pair of timbers.

Shear plates (drawing R) are similar to timber rings, but they are used to make metal-to-timber and concrete-to-timber connections. Both Cleveland Steel and TECO are suppliers of timber rings, shear plates and the grooving tools necessary to install them.

TECO also makes spike grid connectors, which are used to add strength and rigidity to the joints between heavy timbers. Resembling medieval instruments of torture (drawing S), the grids consist of rows of spikes protruding from a malleable iron matrix. They fit between two timbers at a bolted connection, and a threaded compression tool is used to apply enough pressure to the timbers to embed the spikes. For securing a pole to a timber, TECO makes a spike grid that is curved on one side and flat on the other.

Truss clips—If you install nonbearing partitions in a building with trusses overhead, you must not create a rigid connection between the truss and the top plate of the wall. The bottom chord of a truss moves up and down as the loads on it change, and if you don't take its vertical movement into account, the truss can become overloaded.

Steel angles called truss clips can be used to attach a partition wall to a truss while still allowing the truss to move (drawing T, facing page). A slot in the vertical leg of the angle accepts a nail into the lower chord of the truss, anchoring the top of the partition while allowing the chord to move up and down.

Cleveland Steel makes a connector plate, shown in drawing U, that uses the same slot principle to anchor a scissors truss to a wall plate. In this application, the truss wants to move in a horizontal direction, and the slots allow a full inch of movement.

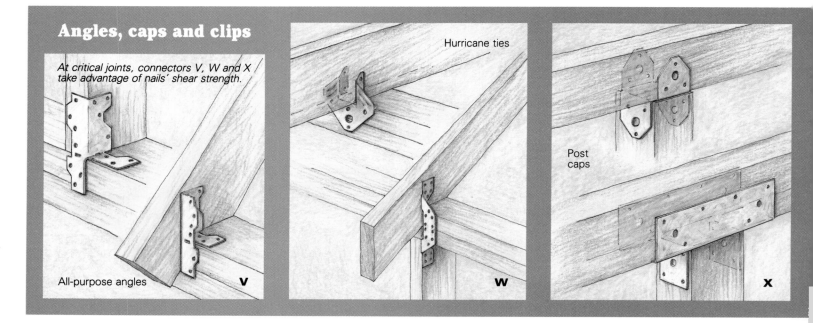

Angles, caps and clips

At critical joints, connectors V, W and X take advantage of nails' shear strength.

Hurricane ties

Post caps

All-purpose angles   V

W

X

Angles, caps and clips—All the companies that manufacture steel connectors make multi-purpose devices that are known as angles, angle clips or reinforcing angles. Their function is to connect butt-joined framing members without toenailing. The more elaborate versions are partially slotted at the fold and have bend lines that allow them to fit a variety of intersections (drawing V, facing page).

Hurricane ties and seismic anchors are another way to avoid toenailing at critical connections. They are folded to wrap around rafters and top plates (drawing W), where they are secured with nails that are working in shear.

Post caps too are designed to get the nails or bolts to work in shear where a post and its beam come together (drawing X). Most are made of 16-ga. galvanized steel, but heavy-duty versions made of 3-ga. painted steel are a standard item from Simpson. Post caps resemble post bases, and in fact some are made to accept a piece of rebar so they can be partially embedded in concrete to become post bases.

By using plywood clips (drawing Y) you can avoid having to use blocking under all the edges of the plywood. The clips will keep the edges of the plywood from seriously deflecting under heavy loads. I use two plywood clips for 16-in. rafter spacing and three for 24-in. rafter spacing. But I don't use them for hot-mopped roofs, because the potential for a little deflection is still there. Instead I'll block under all edges or I'll use T&G plywood.

A slick clip that lets you build a deck without exposed fasteners is made by Philips Manufacturing. It is a galvanized steel angle with prongs on one side that grab the decking on its edge (drawing Z), eliminating rusty nailheads and indented moons from misplaced hammer blows. And since a portion of the clip is sandwiched between adjacent pieces of decking, the clip also acts as a spacer to ensure good drainage.

Specials—Sometimes no commercially available connector can solve a given problem. If I can find something close to what I need in one

of my catalogs, I give the manufacturer a call and ask about modifications. Frequently the company is more than willing to customize a connector for me, and the price has not seemed out of line. (Of course, be sure to ask what the delivery time will be.) To ensure accurate results, supply your fabricator with a full-scale drawing of the special connector that specifies the material to be used, all dimensions, angles and nail or bolt placement.

If you need a truly unusual steel connector, I recommend going to a local sheet-metal or welding shop. A good drawing of the connector is essential. If you are unsure about the loads that the custom connector may have to carry or resist, see if the welding shop can figure them out. Failing that, seek out an architect or an engineer for assistance. □

Bruce Berg is a construction supervisor at Christopherson & Graff, Architects, in Berkeley, Calif.

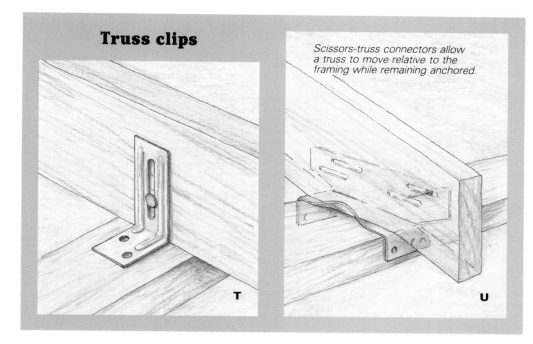

Truss clips

Scissors-truss connectors allow a truss to move relative to the framing while remaining anchored.

T

U

Sources of supply

Cleveland Steel Specialty Co.
14400 South Industrial Ave.
Cleveland, Ohio 44137
(800) 251-8351

Dec-Klip
Philips Manufacturing
460 2nd St.
Lebanon, Ore. 97355
(800) 544-0124

Harlen Metal Products
300 West Carob St.
Compton, Calif. 90220
(213) 774-8383

Heckmann Building Products Inc.
4015 W. Carroll Ave.
Chicago, Ill. 60624-1899
(800) 621-4140

Panel Clip
4203 Shoreline Drive
Earth City, Mo. 63045
(800) 521-9335

Silver Metal Products Inc.
2150 Kitty Hawk Rd.
Livermore, Calif. 94550-9611
(415) 449-4100

Simpson
1450 Doolittle Dr.
P.O. Box 1568
San Leandro, Calif. 94577
(415) 562-7775

TECO
5530 Wisconsin Ave.
Chevy Chase, Md. 20815
(800) 638-8989

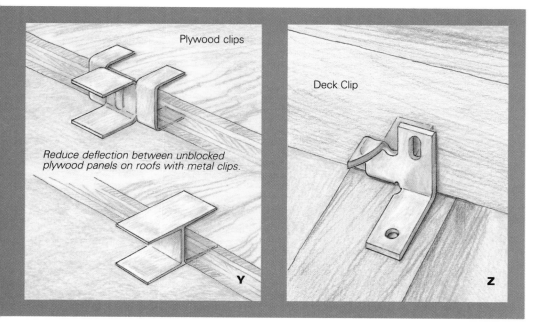

Plywood clips

Reduce deflection between unblocked plywood panels on roofs with metal clips.

Deck Clip

Y

Z

Framing with Trigonometry

Getting to know your scientific calculator will make many construction problems easier

by Edwin Zurawski

When I moved from Hawaii to California a few summers ago to build houses with a friend, I learned something that changed my ideas about how to frame a house. Like most carpenters, I knew how to use rafter tables to find the lengths of common and hip rafters for given slopes and spans. I knew how to step off rafter lengths with the framing square. And I had done my share of direct field measurements, from top plate to ridge. But my friend opened the door to a whole new approach by introducing me to trigonometric framing with a calculator.

Trigonometry is the branch of mathematics that deals with the ratios between the sides of a right triangle (any triangle with one 90° angle). It was first documented around 600 B.C. when Thales of Miletus, a Greek, was visiting with mathematicians in Egypt. He used the principles of trigonometry and the length of his shadow to determine the height of the Pyramids. Fifty years later, Pythagoras developed the equation $a^2 + b^2 = c^2$, describing the relationship between the sides and hypotenuse of a right triangle. But while the Pythagorean theorem is limited to calculating the lengths of a triangle's sides, trigonometry can be used to find degrees of angles as well.

What impressed me most about the trigonometric framing method was how ingenious, yet simple, it is to use. You need to learn only two ratios, tangent and cosine. Every carpenter already knows what tangent is but calls it pitch, or rise/run. Tangent is the ratio between the two sides of a right triangle that form the 90° angle. Cosine is the ratio between one of these sides and the hypotenuse. These ratios are illustrated in drawing A (below left).

Trigonometry has become a more accessible tool for builders because scientific calculators are now so inexpensive that you can get one for $15 or $20. I use a Sharp EL-506P, but any make or model will do as long as it has trigonometric functions (photo below). In the following examples, I've rounded off the numbers to simplify the calculations.

Converting decimals—Since calculators compute in decimals and most building measurements are in feet, inches and fractions of inches, you need to know how to convert from decimal measurements to fractions and vice versa. If you want to convert 6.72 ft. to fractions, for instance, first subtract 6 (the number of whole feet), leaving 0.72 on the calculator screen. Then multiply 0.72 times 12, which converts feet to inches: [.72 × 12 =] 8.64 in. Subtract 8 (the number of whole inches), leaving 0.64 on the screen. Multiply 0.64 times 16 to convert 0.64 in. to sixteenths: [.64 × 16 =] 10.24. Hence 0.64 in. equals approximately $^{10}/_{16}$, or ⅝ in. So 6.72 ft. = 6 ft. 8⅝ in.

To convert fractions to decimals, start with the smallest fraction. For example, if you want to convert 12 ft. 5$^5/_{16}$ in. to decimals, first divide 5 by 16. [5 ÷ 16 =] 0.31 in. Then divide 0.31 in. by 12 to convert it to feet. [.31 ÷ 12 =] 0.026 ft. To convert 5 in. to feet, divide 5 by 12. [5 ÷ 12 =] 0.4167 ft.; [0.4167 + 0.026 =] 0.4427 ft.; and 0.4427 ft. plus the original 12 ft. equals 12.4427 ft.

Calculating common rafters—In order to use trigonometry as a building tool, you have to visualize the right triangles contained within your framework. Once you've done this, if you know the length of one side and the size of one angle, you can use this information to find the remaining dimensions and angles of the triangle. Suppose you're framing a roof with a 6-in-12 pitch on a house that's 30 ft. wide. The length of a common rafter is the hypotenuse of the triangle formed by the ridge, the exterior wall, and the center of the 30-ft. span (drawing B, next page). Here's how to calculate the length of that rafter with trigonometric equations.

On the calculator press [6 ÷ 12 =], and 0.5 appears. That number, rise divided by run, is the trigonometric function known as tangent. You then press [2nd F tan⁻¹] to convert tangent to degrees, and 26.56 appears. The 2nd F key stands for second function and works like the shift key

A. Trigonometric basics

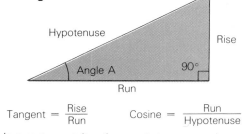

$$\text{Tangent} = \frac{\text{Rise}}{\text{Run}} \qquad \text{Cosine} = \frac{\text{Run}}{\text{Hypotenuse}}$$

Inverse tangent (tan⁻¹) converts tangent to degree. of angle A.

B. Computing common-rafter length

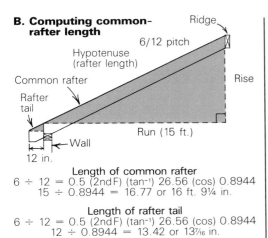

Length of common rafter
$6 \div 12 = 0.5$ (2nd F) $(\tan^{-1})$ 26.56 (cos) 0.8944
$15 \div 0.8944 = 16.77$ or 16 ft. 9¼ in.

Length of rafter tail
$6 \div 12 = 0.5$ (2nd F) $(\tan^{-1})$ 26.56 (cos) 0.8944
$12 \div 0.8944 = 13.42$ or 13⁷⁄₁₆ in.

on a typewriter, allowing some of the keys, tangent in this case, to serve two functions. Tan⁻¹ is called inverse tangent and gives a degree value to a specific tangent. This degree value of tangent can also be used later to make the seat cut of the bird's mouth.

To find the hypotenuse of the triangle, you use run ÷ cosine = hypotenuse. With 26.56 still on the calculator's screen, press [cos] and 0.8944 appears. This is the cosine of 26.56°. Now press [x→m], which stores the cosine of the 6-in-12 triangle in the calculator's memory. Half the span of the 30-ft. house (15 ft.) is the run of the common rafter, so press [15 ÷ RM (recall memory) =] 16.77 ft. (or 16 ft. 9¼ in.). This is the distance from the center of the ridge to the plumb cut at the bird's-mouth cut. The actual rafter of course will have to be shortened by half the thickness of the ridge board. (For more on cutting rafters, see pp. 76-81 and 82-83.)

Rafter tails—If the fascia and soffit detail of your house calls for a 12-in. overhang, then you have to visualize another triangle to calculate the additional rafter length (the hypotenuse of that triangle (drawing B). Begin as you did before, by calculating the cosine of a 6-in-12 pitch. [6 ÷ 12 = 0.5 2nd F tan⁻¹ (26.56) cos (.8944) x→m] to store it. Now divide 12 in. (the length of the overhang) by [RM], press [=], and you get 13.42 in. You would therefore add 13⁷⁄₁₆ in. to the length of your common rafter.

Hip rafters—Since the span of the house is 30 ft., then the level run of the hip rafter is the hypotenuse of a 15-ft. by 15-ft. right triangle (drawing C). On a right triangle with two equal

C. Run of the hip rafter

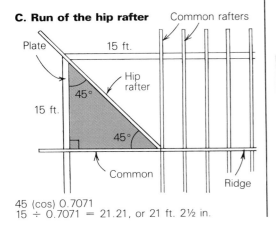

45 (cos) 0.7071
$15 \div 0.7071 = 21.21$, or 21 ft. 2½ in.

sides, an isosceles right triangle, you already know from your high-school geometry that the angle between either of these sides and the hypotenuse is 45°. To find the hypotenuse of a triangle with two equal sides, you simply press [45], then [cos] and 0.7071 appears. Then press [x→m] to store the cosine in the calculator's memory. Now clear the screen and press [15 ÷ RM =] 21.21 ft. This is the level run of the hip.

Hip rafters rise the same distance as common rafters but they have a longer run, since they're cutting diagonally across the building, and therefore have a different pitch. Instead of 12, the unit run of hip rafters is 17, which is the rounded-off length of the hypotenuse of an isosceles right triangle with 12-in. sides (for more on this, see pp. 76-81 and 82-83).

To find the length of the hip rafter in this example (drawing D), you divide 21.21 ft., the run of the hip, by the cosine of a 6-in-17 triangle. First press [6 ÷ 17 =] and 0.3529 appears. Press [2nd F tan⁻¹] and you get 19.44°. Next hit [cos] and 0.943 appears. Press [x→m] and the

D. Length of the hip rafter

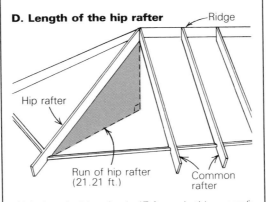

Unit rise of a hip rafter is 17 ft., so in this example the pitch of the hip is 6/17.
$6 \div 17 = 0.353$ (2nd F) $(\tan^{-1})$ 19.44 (cos) 0.943
$21.21 \div 0.943 = 22.49$, or 22 ft. 5⅞ in.

cosine of a 6-in-17 slope is now stored in the calculator's memory. Now clear the screen and press [21.21 ÷ RM =] 22.49 ft. (22 ft. 5⅞ in.) This is the length of the hip rafter.

The hip rafter has to be shortened by a distance equal to half the 45° thickness of the ridge for a single cheek cut or half the 45° thickness of the common rafter for a double cheek cut. Also, the hip must be dropped at the seat cut in order to bring it into alignment with the common rafters on either side of it. Here is the formula for calculating how far to drop the hip rafter: rise divided by 17, multiplied by half the thickness of the hip. As shown in drawing E,

E. Dropping the hip

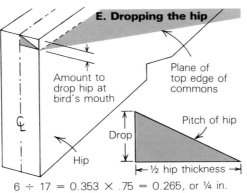

$6 \div 17 = 0.353 \times .75 = 0.265$, or ¼ in.

press [6 ÷ 17 = 0.3529 × .75 (¾ in.—half the thickness of a 1½-in. hip rafter) =] 0.2647 in. I convert 0.2647 in. into sixteenths by multiplying it by 16. [.2647 x 16 =] 4.235 (⁴⁄₁₆) or ¼. You drop the hip ¼ in. for a 6-in-17 slope, if the hip rafter is 1½ in. thick.

Jack rafters—Jack rafters have the same pitch as the common rafters. If you start your jack-rafter layout at the corner of the building, then the on-center spacing of the jack rafters is also the run of the first jack rafter. In drawing F, the first jack is 16 in. o. c. from the corner of the building, and so you find the length of the jack by dividing the run (16 in.) by the cosine of a 6-in-12 pitch.

Here's how to find the length of the first jack rafter. On the calculator, press [6 ÷ 12 =] (0.5

F. Length of jack rafters

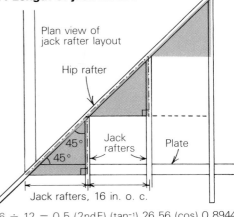

$6 \div 12 = 0.5$ (2nd F) $(\tan^{-1})$ 26.56 (cos) 0.8944
$16 \div 0.8944 = 17.89$, or 17⅞ in.

appears), then [2nd F tan⁻¹] (26.56° appears), then [cos] (0.8944 appears), and finally [x→m]. The cosine of a 6-in-12 pitch is now stored in the calculator's memory. Next clear the screen and press [16 (16 in.) ÷ RM =] 17.89 in. (17⅞ in.). Shorten the jack rafter by half the 45° thickness of the hip rafter (this will be 1¹⁄₁₆ in. for a hip that is 1½ in. thick). Then add 17⅞ in. to each succeeding jack rafter. Note that these jack lengths are the distance along the center of the rafter.

I've described methods for hip and jack calculations as they apply to 45° hips. These methods will work for odd-angle hips, but their application is more complicated. Just remember, you still have to break the framing down visually into right triangles.

Trigonometry and scientific calculators liberate you from rafter tables and help you develop skills that make complex problems seem easy. These methods that I've described to calculate rafter lengths will also work with stairs, or with any construction problem involving angles.

If you spend a little time with tangent and cosine, before long you'll be able to do the calculating at home, so you'll be ready to build when you get to work. Anyone curious about a better way to build and willing to spend $15 for a scientific calculator should give it a try. □

Edwin Zurawski is a general contractor in Concord, Calif.

Putting the Lid On

A primer on production cutting and raising a hip and gable roof

by Don Dunkley

One of the most satisfying events in building a house is the completion of the roof. Some builders borrow from European tradition and nail a pine tree to the peak in celebration. At the least, it is usually the excuse for a party. There are good reasons to celebrate. Framing a roof can be perplexing, physically taxing and sometimes dangerous. However, with thoughtful organization of rafter layout, production rafter-cutting techniques and carefully built scaffolding and bracing to help raise the ridge and rafters, your celebrating doesn't have to come out of a sense of relief.

The best way that I know to share my knowledge of roof framing is to describe the steps involved in building a simple hip and gable roof, like the model roof that is shown in plan, below. This article will cover most of the problems that are encountered in a rectangular building—laying out and assembling common rafters, hips and jacks, along with the ridge, purlins and collar ties.

Preparation—The roof is ready to frame once all the walls are built, plumbed up and braced off. The exterior walls must be lined very straight, because any irregularities in the span will show up on the roof frame. Before you start sorting through your framing stock, study your roof plans carefully. They should show an overhead (plan) view on a scale of ⅛ in. or ¼ in. to 1 ft. They will tell you the type of roof (gable, hip or gambrel), the pitch or slope, the length of overhangs (eave and gable end), the layout of the rafters, their spacing (16 in. on center, 24 in. o.c.), and the sizes of the framing members.

Layout—Job-site layout begins with measuring the span of the building. Always measure from the top (double) plate height. There are usually slight variations between the span shown on the plans, the actual span at the bottom-plate level, and the one at the double plate. Since rafter lengths are calculated down to ¼-in. changes in span, use the double-plate measurement. A 100-ft. tape is the tool for this job.

First, as shown in the photo below, the positions of the rafters must be marked on the top of the double plate. This lets you properly locate the rafters when erecting the ridge. The layout is also necessary to distinguish the positions of the rafters from those of the ceiling-joist layout, which should be placed so they can be used as ties to which the rafters can be nailed. Starting with the hipped end of the roof, lay out the positions of the three king common rafters. Strike a line 10 ft. in from each corner down the length of the building, as well as one midway along the width, and write the letter C (for center) on the plate over each of these lines, which will serve as centers for the king commons. Next, lay out hip-jack rafters on 2-ft. centers from the corner of the building toward the king common rafters.

The common rafters are laid out similarly on the plates, starting at the gable end. I usually mark one side of the rafter position with a line across the top plate. If ceiling joists are also on 2-ft. centers, you don't need to lay them out, because they will be installed beside the rafters. If joists are on 16-in. centers, you would start the layout with the tape held 1½ in. past the end of the top plate. This way, a joist will tie into a rafter every 4 ft.

The ceiling joists that sit on the exterior wall will stick up above the rafters, and can be trimmed along the pitch of the roof after the rafters are up, and before the decking is applied. On the hip, the ceiling joists close to the end wall can't be nailed in place unless you notch them or cheat them off the layout, because the hip will interfere. They should be laid flat on their layouts and installed after the hips and jacks are in place.

Layout tees—The layout tee is a handy tool that lets the builder lay out rafters accurately and quickly. It also helps eliminate steps in rafter-length calculations. Layout tees should be made for the bird's mouth and tail of both

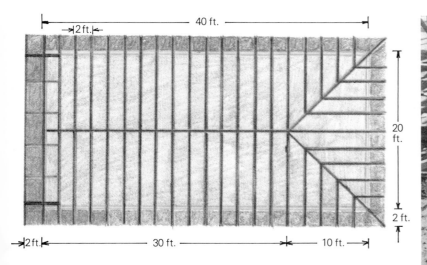

The roof plan of the model, above, shows a gable end using a barge rafter and outriggers for a 2-ft. rake overhang, and a hipped end with a 2-ft. eave. The 2x6 rafters are on 24-in. centers, and the roof pitch is 8-in-12. The span in this case is 20 ft. Right, a carpenter lays out the joists and rafters by walking the plate, something that should be done only after the walls have been plumbed, lined and well braced.

From *Fine Homebuilding* magazine (August 1982) 10:64-69

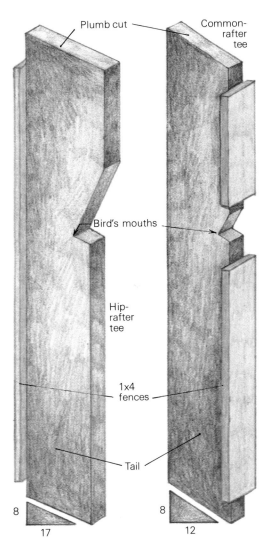

Plumb cut — Common-rafter tee

Bird's mouths

Hip-rafter tee

1x4 fences

Tail

8 / 17   8 / 12

common and hip rafters (drawing, above). There should also be a plumb cut at the top of the tee above the bird's mouth to use as a pattern for the top plumb cut.

Tees should be made of the same width stock as the rafters, so in this case the layout tee for the common rafter should be made from a 3-ft. scrap of 2x6. After you scribe the plumb cut at the top of the tee, move the square about 12 in. down from the top plumb mark and scribe out a bird's mouth. Next, mark the tail cut by measuring along the body of the square for the length of the overhang, which is 2 ft. in this case. Then make a mark and scribe the plumb tail cut, for an 8-in-12 pitch in our example. Cut out the pattern and nail two pieces of 1x4 along the bottom of the tee, one staying clear of the bird's mouth and the other not projecting past the top plumb cut, as shown above. When the tee is being used, the 1x4 fence registers against the bottom edge of the rafter stock for marking out the bird's mouth and plumb cuts.

The hip-rafter tee in this example is cut from a 3-ft. 2x8. On one end, scribe an 8 and 17 plumb mark. Move the square down about 12 in. and scribe the seat cut. Since this is a hip, the seat cut must be dropped (cut more deeply) to bring the top edge of the hip into the same plane as the jacks. To determine the amount of drop, lay the square at the top of the 2x8 on an 8 and 17. Where the 17 mark intersects the top of the lumber, measure down the body of the

square half the thickness of the hip (here ¾ in.) and make a mark. Then measure down from the top of the lumber, perpendicular to the edge, to this mark (⁵⁄₁₆ in.). This is the drop needed. Make the hip seat cut ⁵⁄₁₆ in. deeper than the common seat cuts.

The width of the tail of a hip must equal the width of the common rafter, so the wood past the seat cut must be ripped down from a 2x8 to a 2x6. Measure down from the top edge of the rafter 5½ in. (the actual dimension of a 2x6), mark the length of the rafter tail, and rip the excess 2 in. off the rafter's bottom edge. Since this rip creates a step in the bottom edge, both pieces of 1x4 fence should be nailed to the top of the tee. When using this tee, place it on the top edge of the rafter stock.

The next job is working up a cut list for all the rafters. Count the rafters on your plans, and calculate their lengths. The cuts can then be scribed using the rafter tees, and all the pieces can be cut before beginning the actual installation. This approach requires both confidence and intense concentration, but doing all the cutting first speeds up the process by letting you put your head down and frame without having to stop and figure.

The rafter book—I wouldn't want to be without a rafter book when framing roofs. Mine contains 230,400 rafter lengths for 48 pitches. I can look up any building span under the appropriate pitch, and quickly determine the rafter length and angle of cut. This book saves a lot of labor, and eliminates many costly errors.

Calculating common-rafter lengths—In the rafter book under 8 in 12, the common-rafter table shows that our span of 20 ft. requires a rafter length of 12 ft. ¼ in. from heel cut to plumb cut. This measurement doesn't account for the ridge reduction, because ridge thickness is not a constant. With a 2x ridgeboard, the reduction along a level line is ¾ in. But measured along the rafter edge, ¾ in. measures ⅞ in. on an 8-in-12. Rather than laying out a shortening line on the stock, I subtract the ridge reduction measurement from the rafter-book length to get the corrected rafter length down to the bird's mouth—in this case, 11 ft. 11⅜ in.

The overhang length from the heel cut to the tail cut can be taken off the rafter tee or determined from the rafter book by adding the overhang for each run to the span. A 2-ft. overhang will add 4 ft. to the span. In the rafter book, the 24-ft. span at 8 in 12 reads 14 ft. 5⅛ in. Deduct from that figure the full rafter length of 12 ft. ¼ in. This leaves 2 ft. 4⅞ in. in length from the heel cut to the toe of the tail cut. The overall length of the rafter will be 14 ft. 4¼ in.

Calculating lengths of hips and jacks—Use a large pad of paper to organize your calculations for the hips and their jacks, since they involve a bit of figuring. On the job, keep your building plans clean. Don't scribble math all over them. Using the rafter book, an 8-in-12 hip at a span of 20 ft. is 15 ft. 7⅝ in. A ridge reduction is necessary, and this 45° thickness mea-

sures 1³⁄₁₆ in. along the rafter edge at 8-in-12. This reduces the rafter length to 15 ft. 6⁷⁄₁₆ in. from plumb cut to heel cut.

To find the overhang or tail length, add 4 ft. to the span, just as for the common. The rafter book lists 18 ft. 9⅛ in. for a 24-ft. span. This leaves a rafter tail of 3 ft. 2½ in., and an overall length of 18 ft. 8¹⁵⁄₁₆ in.

To calculate the lengths of the hip jacks, look up jack rafters on 2-ft. centers at 8-in-12. Both the square and the rafter book read 2 ft. 4⅞ in. This distance is the common difference, or how much longer one jack will be than the previous one. This is also the length of the first jack before the deductions. If you are using my system of subtracting the ridge reduction (measured along the edge of the rafter) from the rafter-book length, then subtract 1⅜ in. on an 8-in-12 jack to get 2 ft. 3½ in. from plumb cut to heel cut. Only the first jack needs to be figured for the deduction since the rest will automatically follow, as the common difference is added to each one.

Cutting the rafters—With all the calculations complete, the next step is to lay out and cut the rafters. You can use production techniques that save a lot of time without sacrificing accuracy. I use a rafter bench, an oversize, site-built sawhorse that holds ganged rafter stock up off the ground for easy marking and cutting. I try to set up my benches close to the lumber stack, which should be fairly close to the building. You will be worn out before you start if you have to carry a ton of rafters a great distance.

Stack all the rafters of one type on the bench with their crowns down. The crown is a convex edge seen by sighting down the lumber. Crowns should be placed up in construction to help deflect the load placed on the rafters or joists; they are stacked crown down on the rafter bench so you can scribe cut-lines with the

A chalk line snapped across ganged common rafters marks the heel of the plumb-cut line. As indicated by the layout tee (bottom left) the rafters are stacked with their bottom edges up, and their ends even and square. The layout tee will be used on each rafter to scribe the plumb cut, bird's mouth, and tail cut.

Illustrations: Frances Boynton

Jack rafters stacked on a rafter bench show the common difference of 2 ft. 4⅞ in. on an 8-in-12 pitch using a 24-in. spacing. The diagonal lines indicate the direction of the side cut that will produce pairs of jacks (left and right) for each hip rafter. The bird's mouth and tail will be marked with the common-rafter layout tee.

layout tees. When you stack the rafters on the bench, keep their ends flush so they can be squared up easily with a framing square by drawing a line across their edges. Then use the layout tees to mark the plumb cut at the top, and bird's mouth and tail cuts at the bottom on each of the outside rafters of the stack. Measure the length you have calculated between the plumb cut and the bird's mouth several times, and then connect the marks across the stack with a chalk line (photo facing page, bottom). Lay the first outside rafter down flat on the bench, scribe the plumb, seat and tail cuts with the rafter pattern, and cut them out.

When all the common rafters are cut, they should be dispersed along the exterior walls according to the layout. Before spreading the rafters out, it is a good idea to set a 16d nail at the top plumb cut. This toenail will come in very handy during assembly.

Cut hip rafters using the same procedure, but make double side cuts at the top (a vertical 45° bevel on each side for the plumb cut). These can be made easily on 2x stock with a circular saw set at 45°. For larger timbers or glue-lams, the angle of the top edge of the stock must be laid out, and the cut made with a handsaw.

With the double side cut complete, measure down from the top of the rafter the distance calculated, 15 ft. 6⁷⁄₁₆ in. Mark this length on the center of the rafter's top edge. Slide the tee to this point on the rafter, and scribe the seat cut and heel cut. Mark the rest of the board, scribing along the tail of the pattern. Rip the tail down to the proper width.

Use the common-rafter tee for the jack layout. Group the jacks on the bench according to length—for the model roof, there will be four sets of four. Load the longest first and work down to the shortest set. Only the tail ends can be squared up. Lay the common-rafter tee on this end and scribe the tail and seat cuts on the outside rafter. Then lay out the rafter on the opposite side of the stack and snap lines.

To lay out the plumb cut at the top of the jack, measure the length of common difference—2 ft. 3½ in.—up from the seat-cut line for the shortest set, and add 2 ft. 4⅞ in. progressively to each set of jacks. Square these marks on the top edge of the rafters, and lightly mark two of each set with a 45° line indicating the direction of the angle. Mark the other two with the opposite angle (photo left). The side cut must be laid out this way because the length of a jack is measured from its centerline. Scribe a 45° line in the direction of the light line drawn previously through the center of the plumb-cut line on the top edge of the rafter. Then place the layout tee at the end of the 45° line that intersects the edge of the board farthest down the rafter, and scribe the plumb cut on the face of the rafter. This method creates a slight inaccuracy in the length of the jack on a moderately pitched roof, but it is much faster than marking the precise angle (which can be found in the rafter book or on the square) on the top edge of each rafter.

After cutting all the jack pairs, set them on the roof, paying particular attention to the correct placement of right and left-hand rafters. Drive a 16d nail into the smaller jacks, and hang the head and shank of the nail over the double plate so that the jacks hang along the wall, out of the way but still accessible.

The last roof members to get cut are the ridgeboards and purlins. The 30-ft. ridgeboard on the example here is made from two pieces. Pick straight stock, and cut so the break falls in the middle of a common-rafter layout. The board that includes the hip end of the building should be left long by 6 in., and all cuts should be square.

Assembling the roof—The reward for all the calculating, laying out and cutting is a roof whose members fall right into place once the ridge is up. This is the stage with the largest element of danger, and safety is a primary concern. While nailing joists and laying out the top plate, you'll start to develop "sea legs," gaining confidence in walking around up there. Make sure no loose boards stick out more than a few inches beyond a joist, and keep the top plates between joists free of scraps and nails.

Establishing the ridge height—Before doing anything else, calculate the ridge height to see if you need scaffolding to install it. This is done by multiplying the unit of rise (8 in our example) by the run of the building (10). The bottom of this ridge is 80 in. from the top plate. Ridges 6 ft. or more above the plate need scaffolds. A good scaffold is about 4 ft. lower than the ridge. It must be sturdy, well braced, spanned with sound planks, and running down the center of the building. To make room for the placement of ridge supports and sway bracing, leave a 1-ft. wide space between the scaffold planks.

Raising the gable end and ridge—First, put the tools and materials where you need them. Saws, nails and other tools can be kept handy on a sheet of plywood tacked on the joists. Pull the ridgeboards up on the joists alongside the

scaffolding. You'll need several long 2x4s for braces and legs. Stack them neatly on the joists along with 2x8 bracing for the purlins. To support the gable-end rafters in the initial stage of assembly, nail two uprights to the gable-end wall, perpendicular to the top plate and a foot on each side of the center.

For setting the gable and ridge, you need a crew of four—two carpenters on the scaffold, and one at each end of the span. Starting at the first rafter on the gable end, the carpenters on the outside walls pull up the gable-end rafters, setting the top plumb cut on the scaffold. A small 2x4 block 7½ in. long, the height of the ridgeboard, should be nailed to the plumb cut of one of the rafters. This block temporarily takes the place of the ridgeboard. Make sure the block is flush with the top of the plumb cut. The carpenters on the scaffold pull up the rafters until the seat cuts sit flush on the top plate. The carpenters on the outside walls nail the rafters down, keeping the outside of the gable-end rafter flush with the outside wall. Toenail each rafter to the double plate with two 16d nails on one side and one 16d on the other (back nail). At the plumb-cut end, the rafter with the temporary block must align with the other rafter so that the cuts are nice and tight to the block.

When the gable-end rafters are in position, nail each rafter to the uprights with 16d nails. You'll need to insert a temporary support under the ridge. Measure down from the bottom of the block to the top plate on the gable-end wall to find its length (drawing, below). Nail the leg down to the plate where the 10-ft. center is marked. You'll need another leg under the joint in the ridgeboard, but before you cut it, look for something to set it on. If there isn't a wall directly below the ridge, lay a 2x6 or 2x8 across the joists to carry the leg. After this leg is cut, the block can be removed from the gable end and replaced with the ridgeboard. The carpenter on the other end of the ridgeboard should rest it on the support leg, and scab an 18-in. 2x4 onto the leg and ridge. The scab should stop at least 1 in. below the top of the ridgeboard.

After one end of the ridge is raised, install the common rafter pair that is one layout back from the other end of the first length of the ridgeboard. When you're nailing rafters to the ridge, use three 16d nails to face-nail the first

Supporting the gable end and ridge

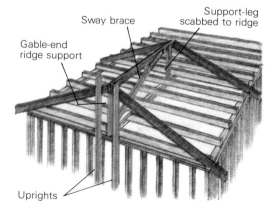

Sway brace

Support-leg
scabbed to ridge

Gable-end
ridge support

Uprights

The king-common rafter that butts the end of the ridgeboard is try-fitted and used to scribe a line for cutting the ridgeboard in place (left). The rafter in the foreground is a king-common that will be nailed at the end of the ridgeboard, perpendicular to the rafter being used for scribing. The skeleton formed by the three king-commons (center left) supports the ridgeboard so the common rafters can be nailed up with frieze blocks. The vertical 2x8 in the foreground is a temporary gable-end brace.

Bottom left: The underside of a hip rafter shows the jack-rafter pairs in position. The upright brace under the hip is placed over a wall. Hips are cut from stock 2 in. wider than common rafters to accommodate the width of jacks cut at compound angles. The added width gives strength for the long span required of hips.

rafter in the pair; then toenail the second. When these rafters are secured, the gable end should be plumbed, and temporarily secured with a swaybrace, a 2x4 with one end cut on a 45° angle, that reaches from the plate (or a 2x8 nailed to the joists above) to the ridge.

Installing hips—The remaining length of ridgeboard is set next. This is easily done by nailing another support leg at the end of the new ridge piece and setting the two king common rafters that define the hip. The third king common, the one that nails to the end of the ridge, is next (photo top left). The hip end of the ridge should extend about 6 in. beyond the king-common layout to allow for final fitting. Do not nail the third common yet, but slide it up to the ridge and scribe the ridge at the plumb cut when the rafter is flush to the top of the ridge and seat cut is up tight (photo center left). Set the rafter down and cut the ridge off, then nail it in place. The resulting frame should be plumb and strong, and ready for the hips.

Raise the hip, pushing its double side cut into the slot at the ridge, and toenail it at the corner and at the ridge. If the hip is spliced, haul the pieces separately on the roof and nail a 2x4 cleat to the bottom edge of the hip at the scarf joint. Position it on the bottom edge so it doesn't interfere with the jacks. Pull a string from the top center of the hip at the ridge, down to the center of the hip at the seat cut. Nail in a temporary upright under the center of the hip and align it with the string. This should eliminate any sag. If it is spliced, cut a leg to fit under the cleat (photo bottom left). Now you can nail the jacks and their frieze blocks.

Jacks, commons and frieze blocks—Start with the smallest jacks and work up. Nail in pairs, to avoid bowing the hip. Then nail the seat cut.

The remaining common rafters can now be filled in, followed by the frieze blocks that go between the rafters at the double plate. The blocks for a 2-ft. o.c. spacing should be cut 22$\frac{7}{16}$ in. and driven tight. Frieze blocks that fit against the hip will have a side cut on one end (photo right). Frieze blocks that are to be nailed perpendicular to the rafters should remain full height. However, if they are to be nailed plumb, they will have to be beveled on the

pitch. This is most easily done on the table saw, but you can do it with a skill saw. In either case, use rafter off-cuts and discarded rafter stock for frieze blocks. For repeated crosscuts, use a radial arm saw; alternatively, you could set up a simple cut-off fixture for your circular saw. The blocks for our example are held to the outside of the top plate, square with the rafter (not plumb) and toenailed flush at the top of the rafter. The next rafter on layout is then pulled up, set in place, nailed at the seat cut and the ridge, and then nailed through the side into the frieze block behind it. Drive two 16d nails for 2x6s; three 16d nails for 2x8s. Whenever a ceiling joist lands next to the rafter, drive three 16d nails through the rafter into the joist.

Purlins—Purlins are required where rafter spans are long. Purlins run the length of the building at the center of the rafter span. They are usually made of the same stock as the ridge, and should be positioned once all the rafters have all been nailed in place. If the commons are 18 ft. or over, it's much easier to handle them if the purlins are installed beforehand. To put up the purlin, first string a dry line across the path of the common rafters to check their sag at the center of their span. Start a purlin at one end, and toenail it into the bottom edge of the rafter, while it's being held by two carpenters. It is held square to the edge of the rafter and perpendicular to the rafter slope. Toenail it to the rafters in several places. Then cut legs (kickers) to fit under the purlin (small photo, facing page). The kickers must sit on the top of a wall, and to avoid deflection should not be placed in the middle of a ceiling-joist span.

Finishing up—Gable roofs are also reinforced with collar ties—horizontal members that connect one rafter in a pair to its opposing member. Collar ties should be no lower than the top one-third of the rafter span. Measure down from the ridge along the slope of the rafter and make a mark about one-third of the way down. Now mark the same distance on the opposite rafter. Hold a 2x4 (or wider board) long enough

This hip rafter has been toenailed in place. The frieze blocks required a single side cut for their intersection with the hip. In cutting the bird's mouth for the hip, the amount of drop had to be calculated. This meant taking a deeper cut so that the top edge of the hip is in the same plane as the other rafters.

The purlin in the foreground (above) is supporting the span of common and hip-jack rafters. Braces positioned at interior walls are perpendicular to the slope of the roof.

Standing on the outriggers (right), a carpenter nails the barge rafter. The frame has been notched for the flat 2x4 outriggers, which are face-nailed to the first rafter inside the gable end, and flat-nailed to the gable-end rafter. The rafters are the top chords of Fink trusses.

to span the two rafters at the marks, and scribe it where it projects past the top of the rafters. Using this as a pattern, cut as many collar ties as you need.

The gable ends must be filled in with gable studs placed 16 in. o.c. Each gable stud fits flush from the outside wall to the underside of the gable rafter. You can make the gable stud fit neatly under the rafter by making square cuts with your saw set on the degree that corresponds with the pitch of the roof. For an 8-in-12 pitch, the corresponding angle is $33\frac{3}{4}°$. You can find the degrees in the rafter book under the pitch of the roof. Gable studs are best cut in sets and, like jack rafters, they advance by a common difference.

The example shows a rake of 2 ft., with a barge rafter and outriggers. Unlike the fly rafter and ladder system shown in the glossary (p. 83), a barge rafter usually isn't reduced for the ridge; it butts its mate directly in front of the end of the ridge board. The outriggers support the barge overhang. They are typically 2x4s, 4 ft. o.c. from the ridge down, extending from the barge rafter across the gable-end rafter and beyond one rafter bay. The outriggers are notched into the gable rafter, laid in flat and face-nailed to the common rafter in back, as shown in the photo at right.

To put in outriggers, first lay out the top of the gable rafter 4 ft. o.c., starting from the ridge. The layout should be for flat 2x4s ($3\frac{1}{2}$ in. wide). Then notch the layout marks with several $1\frac{1}{2}$-in. deep saw kerfs and a few quick blows from your hammer. Make these cuts down on the rafter bench. Let the outriggers run long and cut them along a chalked line once they are up to ensure a straight line for nailing the barge rafter. □

Don Dunkley is a carpenter and contractor in Sacramento, Calif.

Roof Framing Simplified

This direct approach involves full-size layouts and stringing rafter lines

by Tom Law

There isn't a cut in roof framing that can't be calculated given a sharp pencil, a framing square and a head for math. But my 20 years in the trade have taught me that in some cases, the theoretical calculation of rafter angles and lengths is slower and leaves more room for error. While I think that it's important to understand the geometry of roof framing, the empirical method can save time and frustration, and contribute to your understanding of the process. I'm better at solving problems when I can grasp them—literally.

When I'm cutting a complicated roof and things get foggy, I use two techniques to help me produce rafters that fit the first time. I chalk lines on the plywood subfloor to represent a rafter pair in relation to its plates and ridge. This two-dimensional diagram is laid out full size. Pattern rafters can be tested right there on the job site. The other method I use is to deal directly with the components involved by getting up on the roof and measuring the relationships between the rafter to be cut and the existing plate and ridge with string and sliding bevel.

I used this method several years ago when I built a Y-shaped house. One wing was for the bedrooms and the other contained the kitchen, dining room and family room. The stem of the Y was the living room. The roof over the wings called for trusses, but the living-room rafters were exposed. The problem was in framing the intersection of the three roofs. These beams were big, long and expensive, and all the cuts would show. Had the house been a T shape, the valley rafters would have been a textbook case, and I could have found the information I needed in the rafter tables on my framing square. But since the intersection was 120° and not 90°, I had to find the angles by calculation or direct observation. I chose the latter.

Full-size layout—After setting the trusses, I chalked a full-size layout of the living-room common rafters and ridge on the subfloor below. The ridge beam was a 4x14, and the common rafters were 4x8s on 4-ft. centers. I decided to tackle the easiest steps first. This gave me time to think about the problem while reducing the parts in the puzzle.

The ridge beam went up first. I found its height by measuring on the full-size layout. The common rafters were then cut using patterns made from the layout. I nailed these in place

Tom Law is a builder in Davidsonville, Md.

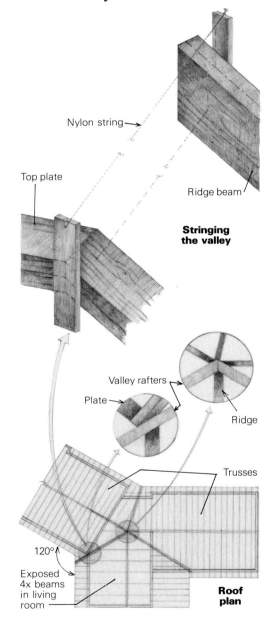

Nylon string

Top plate

Ridge beam

Stringing the valley

Valley rafters

Plate

Ridge

Trusses

120°

Exposed 4x beams in living room

Roof plan

starting at the outside wall, working toward the junction of the Y.

The valley rafter was next. Its location is shown above. Valley rafters are usually heftier than common rafters because they have a longer span, and have to carry the additional weight of the valley jacks. In this case, the valley rafter was a 4x10. Because it's deeper, to achieve the same height above the plate and ridge, its seat and ridge cuts also differed from those of the common rafters. I found it faster and easier to measure the actual distances than

to calculate imaginary ones. This way I could find the length of the rafter and the angles of the cuts without guesswork or error. This took me back up on the roof armed with a ball of nylon string, a sliding bevel and a level.

String lines—First, I tacked a scrap piece of wood vertically on the opposite side of the ridge beam from its intersection with the valley rafter. Then I tacked another block on the outside of the top plate where the valley rafter would sit. Between these sticks, I stretched nylon string at the height of the top of the rafter and along its imaginary center line. The string made it easy to visualize the actual rafter in place. For reassurance, I sighted across the rafter tops from the outside wall to check the alignment. I used the sliding bevel and level to find the angle of all the cuts, being careful not to distort the string.

Before I transferred the angles to the rafter stock, I made two templates (called layout tees) for marking out the ridge cut and rafter seats. I made one for the bottom and one for the top of the valley rafter. With some adjustments, they fit when the center line drawn on the pattern was in line with the string representing the center of the rafter. The tees also allow you to test-fit the bird's mouth to the plate before you carve up costly rafter stock. With these pattern pieces tacked in place, I measured the length of the valley directly. Then I transferred this length and the angles on the tees to the valley-rafter stock, and cut it to its finished dimensions. It fit perfectly the first time.

With the valley rafter in place, I turned to the valley jacks. First I made another pattern, this time of the ridge cut of the common rafter. I tacked it on the layout mark on the ridge and stretched the nylon line from it to the valley rafter, being careful to keep it exactly parallel to the common rafters. With the line simulating the top center line of the longest jack, I used the sliding bevel to find the angles of the plumb and side cuts. This time I transferred the angles directly onto the stock and cut it with a handsaw. Each shorter jack was worked in the same way, using the nylon line to find the location and length; the angles remained constant.

Laying out rafters with a framing square is something I do a lot. However, in situations that call for unusual intersections with compound angles, I spend my time dealing directly with the problem. This reduces confusion and allows me to concentrate on the work. □

From *Fine Homebuilding* magazine (August 1982) 10:62-63

A Glossary of Roofing Terms

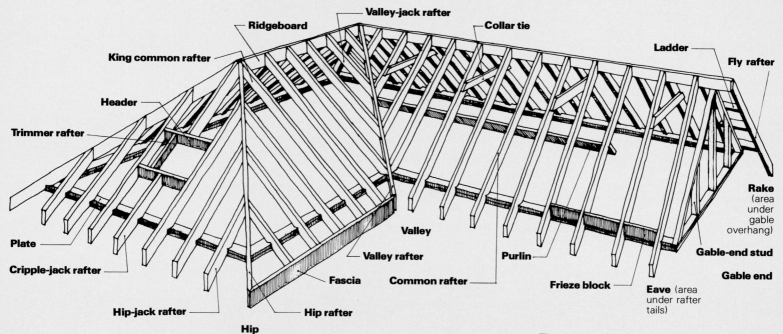

The names of the roof members (above), and the rafter terms (defined below) vary according to geographical region and roof style. For an explanation of how hip and gable roofs are framed, see the preceding article.

Span—the horizontal distance between the outside edges of the top plates.

Rise—the vertical distance measured from the wall's top plate to the intersection of the pitch line and the center of the ridge.

Run—the horizontal distance between the outside edge of the top plate and the center of the ridge; in most cases, half the span.

Slope—a measurement of the incline of a roof, the ratio of rise to run. It is typically expressed using 12 as the constant run.

Pitch—has become synonymous with slope in modern trade parlance. It is actually the ratio of the rise to the span. A roof with a 24-ft. span and a rise of 8 ft. has a 1-to-3 pitch. Its slope is 8 in 12. Two ways of saying the same thing.

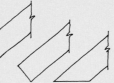

Unit rise—the number of inches of rise per foot of run.

Unit run—this distance is always 12 in.

Common difference—the difference between the length of a jack rafter and its nearest neighboring jack on a regular hip or valley when they are spaced evenly. This is also the same measurement as the length of the first, or shortest, jack.

Rafter pattern—a full-scale rafter template used to mark the other rafters for cutting. It can be tried in place for fit before cutting all the rafters.

Layout tee—a short template cut from the same stock as the rafters and used for scribing repetitive plumb cuts, tail cuts and bird's mouths.

Tail—the part of a rafter that extends beyond the heel cut of a bird's mouth to form the overhang or eave.

Pitch line—an imaginary line, also called the **measuring line**, that runs parallel to the rafter edges at the height of the full depth of the heel cut on the bird's mouth. In common practice, rafters are measured along their bottom edge.

Theoretical length—the length of a rafter without making allowances for the tail or ridge reduction. Also called the **unadjusted length**.

Bird's mouth—also called a **rafter seat**. It is the notch cut in a rafter that lets it sit on the double plate. It is formed by the plumb heel cut and the seat cut, which is a level line.

Plumb cut—any cut that is vertical when the rafter is in position on the roof. Also used as a reference to the top cut on a rafter where it meets the ridgeboard.

Level cut—any cut that is horizontal when the rafter is in position on the roof.

Tail cut—the cut at the outer end of the rafter. If cut at the outside edge of the double plate, it is a flush cut. All the other traditional tail cuts let the rafter overhang the plates—**heel cut** (level), **plumb cut** (vertical), **square cut** (perpendicular to the length of the rafter) or **combination** level and plumb cuts.

Side cut—also called a **cheek cut**, is the compound angle required for the proper fitting of roof members that meet in an intersection of less than 90°, and other than level. This applies to jacks that connect with hips and valleys.

Ridge reduction—rafter lengths are calculated to the center of the ridge of the roof. This doesn't take into account the thickness of the ridgeboard. This allowance reduces the theoretical length of the rafter by one-half the thickness of the ridgeboard. The layout line drawn parallel to the plumb cut that represents this allowance is called the **shortening line**.

Dropping a hip—the amount by which the bird's mouth on a hip rafter must be deepened to allow the top of the rafter to lie in the same plane as the jack and common rafters. This ensures that the roof sheathing will nail flat without having to bevel the top edges of the hip, a process known as **backing**. —*P.S.*

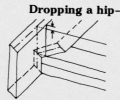
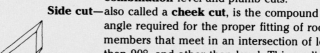

Illustrations: Frances Boynton

Dummy Rafter Tails

A useful technique for old-house remodelers and those who build alone

by Bob Syvanen

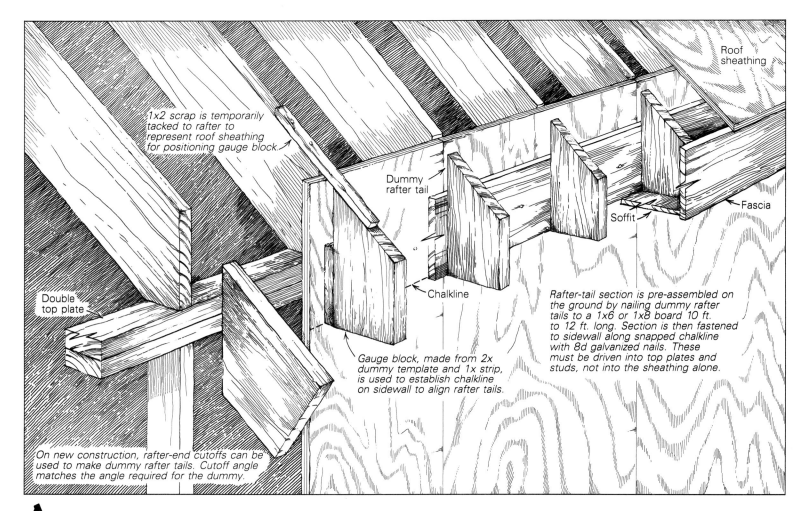

1x2 scrap is temporarily tacked to rafter to represent roof sheathing for positioning gauge block.

Roof sheathing

Dummy rafter tail

Fascia

Soffit

Chalkline

Double top plate

Gauge block, made from 2x dummy template and 1x strip, is used to establish chalkline on sidewall to align rafter tails.

Rafter-tail section is pre-assembled on the ground by nailing dummy rafter tails to a 1x6 or 1x8 board 10 ft. to 12 ft. long. Section is then fastened to sidewall along snapped chalkline with 8d galvanized nails. These must be driven into top plates and studs, not into the sheathing alone.

On new construction, rafter-end cutoffs can be used to make dummy rafter tails. Cutoff angle matches the angle required for the dummy.

A leaky roof or a backed-up gutter often leads to eaves and rafter tails that are badly deteriorated. On older houses, there's simply nothing solid to nail new fascia and soffit trim to. To deal with this problem, I cut the old rafter tails back flush with the siding, nail dummy 2x6 block rafter tails to a 1x board, and nail this in turn through the sheathing to the top plate, the cut-back rafter ends, or to both.

I've found that this dummy rafter-tail system also makes it easy for one person to nail up rafters on a new house. This procedure also eliminates the need to cut and install frieze blocks between rafters atop the plates. I extend the plywood siding 4 in. to 6 in. above the top plate, and it holds the bottom end of the rafter in place while I nail the ridge end. This sometimes lets me get away with using standard-length (20-ft. or less) rafter stock instead of having to special-order something longer. In these cases, I make up the blocks from the scrap lumber that's all over the job site. When length isn't a problem, I use cutoffs from the rafters themselves for the dummy tails. They've already got the correct angle cut at one end, so I have only to cut them to length and width.

Step by step—Not all framing material is a consistent width, so if I'm making up my dummy rafter tails from 2x6 stock, I begin by running the block stock through a table saw to make sure all the pieces are the same width. I cut as many rafter-tail blocks as there are rafters, then nail them to a 1x6 or 1x8 board. On a house 36 ft. long, I make up three 12-ft. sections for each side—longer lengths are too unwieldy.

To position the block strip, I use a gauge block made up of a rafter-tail block with a piece of 1x stock nailed to the back. I align the gauge block with the top of the rafter (on new construction, a 1x2 temporarily tacked to the rafter top to represent roof sheathing helps here), hold it against the sidewall sheathing, and mark the bottom. It's best to mark both ends of the wall and the middle, then snap a line the full length. This will show up any variation that might have resulted from the guide marks having been taken off a high or low rafter. To get a good long-line snap, I thumb the middle of the line and snap each side.

Once the dummy rafter tails are in place, I install the roof sheathing over them, tying the whole business together. The end result is rafter tails that line up straight as string. □

Bob Syvanen is a consulting editor of Fine Homebuilding *magazine.*

From *Fine Homebuilding* magazine (June 1985) 27:43
Drawing: Elizabeth Eaton

Roof Framing Revisited

A graphic way to lay out rafters without using tables

by Scott McBride

When architect Stephen Tilly presented me with the drawings for a house he planned to build in Dobbs Ferry, N. Y., I flipped through the pages one at a time. I stopped when I got to the roof-framing plan. In addition to the roof's steep pitch (12-in-12) and substantial height (more than 40 ft. in some places), I would have to deal with a plan that included few corners of 90°, roof intersections of different pitches and a portion of a cone.

Irregular plan—When adjoining roof planes rise at the same pitch, their intersection (either a hip or valley) bisects the angle formed by their plates on the plan. For instance, a square outside corner on a plan calls for a hip whose run bisects the plates at 45°, as shown in the drawing below. This is why the skillsaw is set to a 45° bevel in most cases to cut hips. Similarly, square inside corners produce valleys lying at 45° to their adjoining ridges.

When plates join at an angle other than 90°, things get a bit more complicated. The Dobbs Ferry house contains a number of these irregular situations, including a large bay that angles off the master bedroom at 45°, and a wing that joins the main body of the house at 60°.

Theory—Conventional methods of roof framing involve the extensive use of rafter squares, but I decided to take a different approach. My study began by reading *Roof Framing* (Sterling Publishing Co., Two Park Ave., New York, N. Y. 10016) and *The Steel Square* (Drake Publishers, 381 Park. Ave. South, New York, N. Y. 10016), both by H. H. Siegele. These books are a hap-

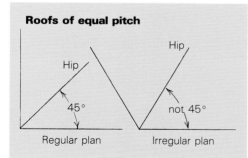

Roofs of equal pitch

Hip

Hip

45°

not 45°

Regular plan

Irregular plan

An irregular plan. **With intersecting roof planes of the same pitch, the hip or valley bisects the angle formed by their plates in plan (drawing, above). On this house, right, the main wings join at 120°. As a result, the valley is an irregular one, since the angle at which it bisects the plates in plan is other than 45°.**

From *Fine Homebuilding* magazine (August 1985) 28:31-37
Photos: Stephen Tilly

hazardly arranged series of magazine articles from the 1940s, and the illustrations are poorly reproduced in the current editions. Nevertheless, they contain the necessary applied geometry that is the essence of roof framing.

Understanding Siegele and the subject of irregular framing in general isn't easy. I spent hours rereading a page; in some cases, even a single cryptic paragraph. But when, after following his convoluted instructions, that weird cut I made on an 18-ft. double 2x10 valley rafter fit perfectly, it all seemed worthwhile.

Preparation—I began work on the house by making a ½-in. = 1-ft. scale model of the entire house frame (photo right). Lofted directly from photographically enlarged working drawings, it allowed architect Tilly and me to analyze the downward transfer of the roof load, and to visualize the relationship of one rafter to another in three dimensions. Finally, the model proved helpful with the materials takeoff, since determining the rough length of hip, valley and jack rafters from a plan alone is difficult on a complex roof.

Once the last double top plate on the second story was nailed down, I began to deal with the roof in earnest. I drew two roof plans to use in the actual framing process. The first is a schematic (drawing, below) showing the position of plates, ridges, rafters, etc. This plan, like the model, gave me a general idea of how the roof

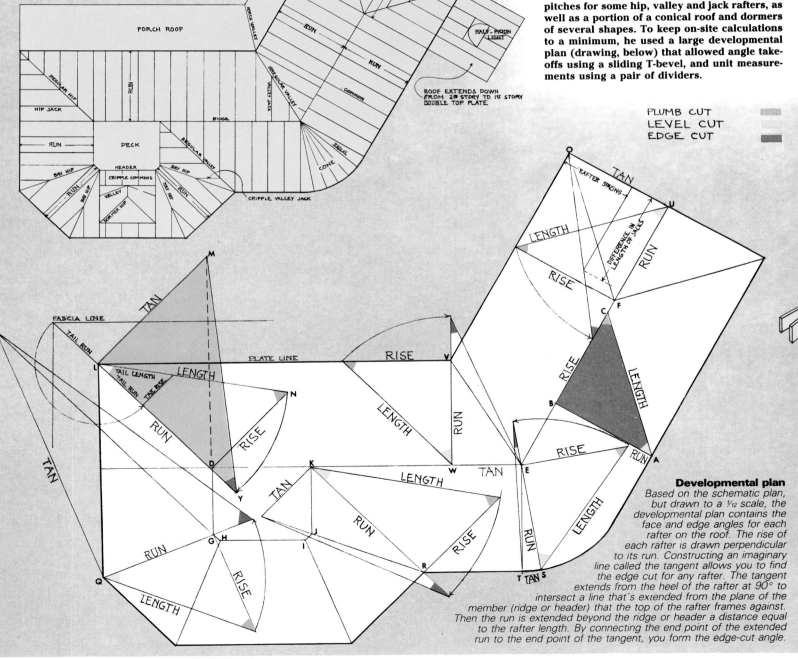

The structural model, above, shows just how complicated the roof is. The author had to deal with both an irregular plan and irregular pitches for some hip, valley and jack rafters, as well as a portion of a conical roof and dormers of several shapes. To keep on-site calculations to a minimum, he used a large developmental plan (drawing, below) that allowed angle take-offs using a sliding T-bevel, and unit measurements using a pair of dividers.

Schematic roof plan
This plan was drawn to the same scale as the blueprints, and its purpose was to give the author an overall idea of how the entire roof went together—the exact number and location of all the plates, ridges, headers and rafters, both regular and irregular.

Developmental plan
Based on the schematic plan, but drawn to a 1/12 scale, the developmental plan contains the face and edge angles for each rafter on the roof. The rise of each rafter is drawn perpendicular to its run. Constructing an imaginary line called the tangent allows you to find the edge cut for any rafter. The tangent extends from the heel of the rafter at 90° to intersect a line that's extended from the plane of the member (ridge or header) that the top of the rafter frames against. Then the run is extended beyond the ridge or header a distance equal to the rafter length. By connecting the end point of the extended run to the end point of the tangent, you form the edge-cut angle.

Drawings, except where noted: Scott McBride

would work, and I drew it to the same scale as the prints. On this plan, the common rafters are always oriented perpendicular to the plates, and the run, or horizontal distance traveled by the rafters, is the same on both sides of the roof.

I drew the second plan (drawing, bottom of facing page) at 1 in. = 1 ft.—four times as large as the first plan. This developmental plan contains the face and edge angles of each common, hip, valley and jack on the roof, making it possible for me to transfer them from the paper right onto the rafter stock with a sliding T-bevel. (For definitions of roof members and roofing cuts, see the drawing below.)

Also, the graphic constructions on this plan provide unit rafter lengths equal to 1/12 the extended rafter lengths since I used a 1-in. = 1-ft. scale. To get full-scale lengths, I adjust a large pair of dividers to the unit length on the plan and step off this distance twelve times along the measuring line on the rafter stock. This gives me the *extended length* (or unadjusted length) of each rafter. The extended length is the distance from the center of the ridge to the outside edge of the top plate. To find the actual length of a common rafter, you must deduct one-half the thickness of the ridge and add the necessary amount to the tail for overhang at the eaves. I like this graphic system because it avoids math entirely—no rafter tables, no Pythagorean theorem. Also, the dividers are handier and more accurate for stepping off than the steel square.

Using the plan with commons—To draw the plan, I began with the development of a common rafter, which is seen in plan as line segment AB in the developmental drawing (facing page, bottom). The distance from A to B is the *run* of the rafter. It represents the distance across the attic floor from the outside edge of the plate to directly below the center of the ridge, and it forms the base of a right triangle. This triangle carries the geometric information you need to know about a rafter to cut it. The other leg of the triangle, line BC, is the *rise*. In actual three-dimensional space the rise goes from point B on the attic floor to point C, directly above B on top of the ridge. The hypotenuse of this triangle, line AC, is the unadjusted length of the rafter. The angle opposite the run is the plumb-cut angle, and the angle opposite the rise is the seat-cut angle.

If you are thinking in three dimensions, you're probably confused. You can see that right triangle ABC is lying flat instead of standing in three dimensions as it would on a model. It is literally pushed over on one side, using AB as a kind of hinge. This is the key to understanding the way this developmental drawing expresses spatial relations using a flat piece of paper. The constructions for all of the rafters in the drawing employ the same basic transposition by bringing the rise and the rafter length down onto the paper, along with the run.

From paper to rafter stock—Cutting a common rafter from the developmental drawing requires taking the information graphically represented in the plan and applying it full scale to the rafter stock. For the common rafters represented by line AC, use a sliding T-bevel to transfer angle BCA onto the rafter stock. This will

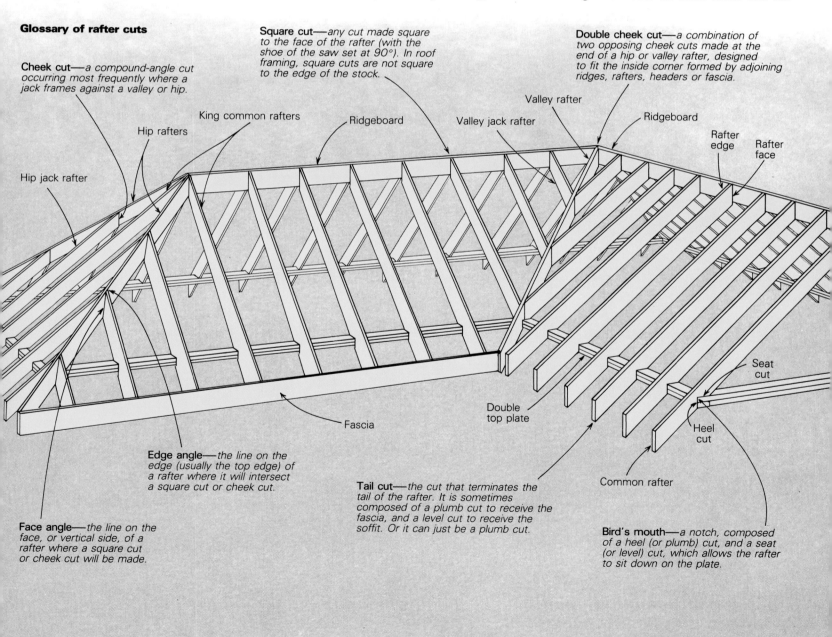

Glossary of rafter cuts

Square cut—*any cut made square to the face of the rafter (with the shoe of the saw set at 90°). In roof framing, square cuts are not square to the edge of the stock.*

Double cheek cut—*a combination of two opposing cheek cuts made at the end of a hip or valley rafter, designed to fit the inside corner formed by adjoining ridges, rafters, headers or fascia.*

Cheek cut—*a compound-angle cut occurring most frequently where a jack frames against a valley or hip.*

Hip jack rafter

Hip rafters

King common rafters

Ridgeboard

Valley jack rafter

Valley rafter

Ridgeboard

Rafter edge

Rafter face

Seat cut

Double top plate

Fascia

Edge angle—*the line on the edge (usually the top edge) of a rafter where it will intersect a square cut or cheek cut.*

Tail cut—*the cut that terminates the tail of the rafter. It is sometimes composed of a plumb cut to receive the fascia, and a level cut to receive the soffit. Or it can just be a plumb cut.*

Heel cut

Common rafter

Face angle—*the line on the face, or vertical side, of a rafter where a square cut or cheek cut will be made.*

Bird's mouth—*a notch, composed of a heel (or plumb) cut, and a seat (or level) cut, which allows the rafter to sit down on the plate.*

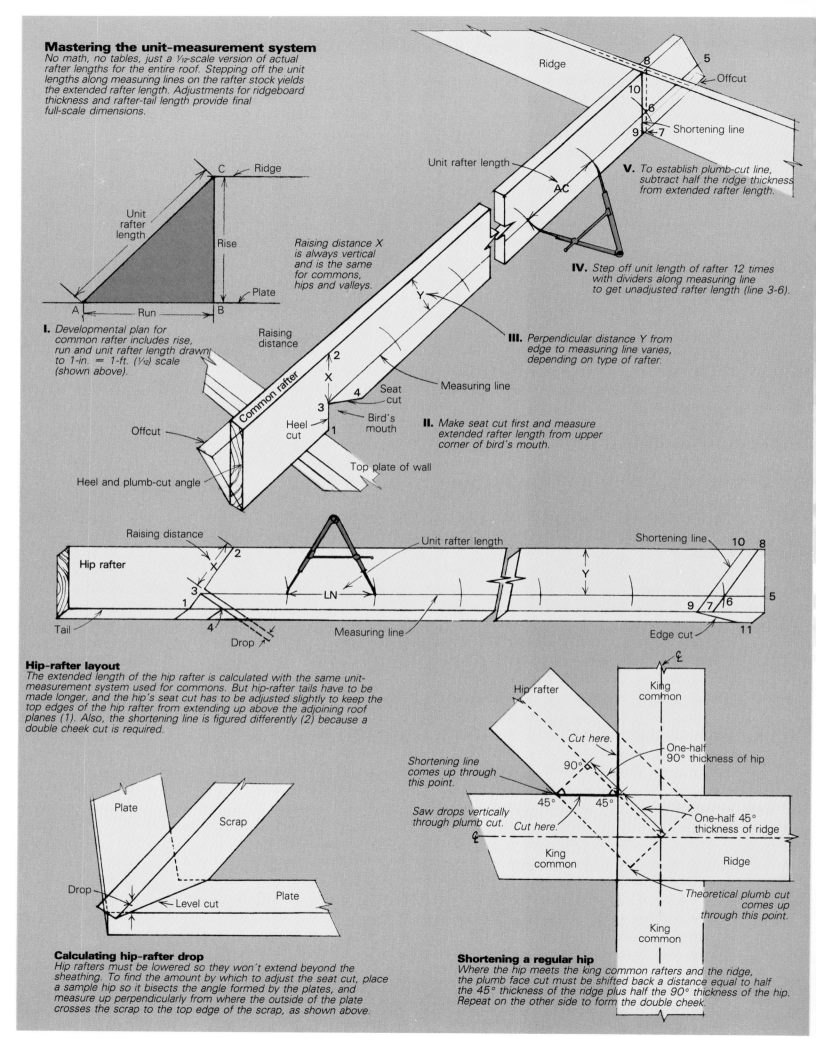

Mastering the unit-measurement system

No math, no tables, just a ¹⁄₁₂-scale version of actual rafter lengths for the entire roof. Stepping off the unit lengths along measuring lines on the rafter stock yields the extended rafter length. Adjustments for ridgeboard thickness and rafter-tail length provide final full-scale dimensions.

Ridge

C — Ridge

Unit rafter length

Rise

Plate

A — Run — B

Raising distance X is always vertical and is the same for commons, hips and valleys.

I. Developmental plan for common rafter includes rise, run and unit rafter length drawn to 1-in. = 1-ft. (¹⁄₁₂) scale (shown above).

Raising distance

Common rafter

Offcut

Heel and plumb-cut angle

2
X
3   4   Seat cut
Heel cut   1   Bird's mouth

Measuring line

Top plate of wall

II. Make seat cut first and measure extended rafter length from upper corner of bird's mouth.

Unit rafter length

AC

Offcut
Shortening line
8   5
10
6
9   7

V. To establish plumb-cut line, subtract half the ridge thickness from extended rafter length.

IV. Step off unit length of rafter 12 times with dividers along measuring line to get unadjusted rafter length (line 3-6).

III. Perpendicular distance Y from edge to measuring line varies, depending on type of rafter.

Raising distance

Hip rafter

Unit rafter length

Shortening line   10   8

2
X
3
1
4
Drop
Tail
Y
LN
Measuring line

9   7   6
5
11
Edge cut

Hip-rafter layout

The extended length of the hip rafter is calculated with the same unit-measurement system used for commons. But hip-rafter tails have to be made longer, and the hip's seat cut has to be adjusted slightly to keep the top edges of the hip rafter from extending up above the adjoining roof planes (1). Also, the shortening line is figured differently (2) because a double cheek cut is required.

Plate

Scrap

Drop

Level cut

Plate

Plate

Hip rafter

King common

Cut here.

Shortening line comes up through this point.

Saw drops vertically through plumb cut.   Cut here.

90°

45°   45°

One-half 90° thickness of hip

One-half 45° thickness of ridge

King common

King common

Ridge

Theoretical plumb cut comes up through this point.

Calculating hip-rafter drop

Hip rafters must be lowered so they won't extend beyond the sheathing. To find the amount by which to adjust the seat cut, place a sample hip so it bisects the angle formed by the plates, and measure up perpendicularly from where the outside of the plate crosses the scrap to the top edge of the scrap, as shown above.

Shortening a regular hip

Where the hip meets the king common rafters and the ridge, the plumb face cut must be shifted back a distance equal to half the 45° thickness of the ridge plus half the 90° thickness of the hip. Repeat on the other side to form the double cheek.

give you a plumb line. Use it first to establish the heel cut 1-3 near the bottom of the rafter (drawing, facing page, top), making sure to leave enough stock below this line for the rafter tail.

Now, holding the blade of the T-bevel perpendicular to the heel cut, use angle BAC against the edge of the rafter to scribe the horizontal seat cut 3-4. I make the seat cut 3½ in. to equal the width of the rafter plate. Cutting away triangle 1-3-4 will establish the bird's mouth.

To establish a length on the rafter, you have to scribe a line to measure along. Starting at the inside corner of the bird's mouth (point 3), and proceeding parallel to the edge of the rafter, scribe the *measuring line* 3-5. Going back to the plan, set the dividers to the unit length of the rafter AC. Starting from point 3 again, step off this increment 12 times along the measuring line. This will bring you to point 6. Draw a plumb line 7-8 (it will be parallel to line 1-2) through this point. This represents the imaginary line down through the exact center of the ridge, and the distance from 3 to 6 is the unadjusted length of the rafter.

To make the actual face cut for the rafter where it bears against the ridge, draw line 9-10 (known as the *shortening line*) parallel to line 7-8. The shortening line is offset from the theoretical plumb cut by a horizontal distance equal to one-half the thickness of the ridge—in this case ¾ in.

Note that the top edge of the rafter is offset from the measuring line by the vertical distance 3-2 (equal to 6-8). This is what I call the raising distance (X). It represents the difference between the level at which the roof is calculated and laid out, and the tops of the actual rafters. Measuring lines on all rafters are offset from their respective edges by this same raising distance, as measured vertically along their respective plumb cuts. If all the planes of a roof are to meet smoothly, then this measurement must remain constant from rafter to rafter.

Raising the roof—After cutting all the commons rafters, I cut temporary posts to support the ridge. I erected these at points D, E, and F (developmental drawing, p. 86) and toenailed ridge boards DE and EF in from above. Headers DG, GH, HI, IJ and JK were propped up at the same elevation as the ridge. There was a reason DG and KJ were not permitted to intersect directly with HI in a square corner: given the dimensions of the floor plan below, to have done so would have made the run of the canted bay roofs less than the run of the adjoining roofs. Since the rise of all roofs here is the same, and pitch is the function of run and rise, giving the deck a square corner would have given the bay roofs a slightly different pitch than their neighbors. This would have combined an irregular pitch condition with the existing irregular plan. Have mercy!

Regular hips—After spiking in all the commons, I proceeded to draw the regular hips (DL, OF, and PF) starting with DL on the developmental plan. I first drew the rise DN, perpendicular to the run (DL). Then I connected L to N to produce the unit length of the rafter as seen in profile. Note that the rafter length of the hip is

greater than that of the common, because its run is greater, even though the rise is the same.

In much the same way as I did with the common rafters, I used the T-bevel set to angle DNL from the plan to strike the heel cut on the rafter stock, again making sure to leave enough length on the tail for eventual trimming at the fascia (middle drawing, facing page). Hips and valleys require longer tails than commons and jacks, because although the rise of the hip tail is the same as the rise of the common tail, the run of the hip tail is greater. To calculate the length required for the hip tail, I drew in the fascia lines to scale on the plan parallel to the plate lines.

Extending the run of the hip until it intersects the fascia line produces the run of the tail. Set the dividers to this increment and swing them around 180° to a point along the hip run. Squaring up from this mark will establish the rise, and its intersection with the hip length establishes the end point for the tail length. Remember, though, this is a unit length, and will require being stepped off 12 times on the rafter.

The next task is to scribe the rafter stock with a measuring line from the inside corner of the seat cut to the top end of the rafter. It's parallel to the top edge of the rafter and offset from it by the same vertical raising distance established on the common rafters. To complete the bird's mouth on this hip, I drew a line through the intersection of the measuring line and the heel cut, using angle DLN on the T-bevel. This produced the seat cut. Well, almost.

Dropping the hip—A hip seat cut has to be *dropped*, which is the procedure of lowering the entire hip slightly in elevation by making a somewhat deeper seat cut. If the hip were not dropped, the corners on its top edge would protrude slightly beyond the adjoining roof planes, interfering with the installation of the sheathing.

To determine the amount of drop necessary, I took a piece of scrap the same thickness as the hip (drawing, facing page, bottom left) and made a level cut through it, using angle DLN. I placed this so that it bisected the 90° angle formed by the intersection of the adjoining plates. I then squared up from the level cut at the point where the outside of the plate crossed the face of the scrap. The distance from this point to the top edge of the scrap was the amount of drop. Accordingly, I moved up the theoretical seat cut a vertical distance equal to the drop, to arrive at the actual seat cut (middle drawing, facing page).

When stepping off a hip for length, you start from point 3 and repeat the unit hip-rafter length LN (obtained from the plan) 12 times. Using angle DNL, I struck a plumb line through the point 6. This is line 7-8, and represents the center of the ridge. Shifting this plumb line back horizontally a distance equal to half the 45° thickness of the ridge (1¹⁄₁₆ in. for a 2x ridge) plus half the 90° thickness of the hip (¾ in. for a 2x hip) established the actual plumb face cut (shortening line), 9-10. The bottom right drawing on the facing page demonstrates why these reductions are necessary by showing hip plumb cuts.

To complete the hip rafter, the shortening line has to be laid out in the same relative position

on both sides of stock, and cut with a skillsaw set at a 45° bevel from each side. This produces a *double cheek cut* (see the drawing on p. 87), which nuzzles into the corner formed by the adjoining faces of the king common rafters.

Irregular hips and valleys—After completing the regular hips, I was ready for a bigger challenge—the irregular parts of the plan. The easiest of the irregular rafters to cut (for reasons I will explain later) was the group of bay hip rafters typified by QG, so I started with those.

All cheek cuts, regular and irregular, fall into one of two categories: those that can be cut with a skillsaw (with a bevel of 45° or greater) and those that cannot (with a bevel less than 45°). This difference is determined by the angle formed between the run of the given rafter, and the ridge, header or other rafter against which it frames. For instance, the irregular hip rafter QG frames against header GH at a 67½° angle (112½° if measured from the opposite side of the rafter). As a result, it can be easily cut with a skillsaw, and in a sense is no more difficult to cut than a regular hip.

The first step is to develop the unit rise and the unit length of the rafter on the plan, using the same graphic procedure as for common rafters and the regular hips. Again, you will need to draw a measuring line parallel to the edge of the rafter stock, offsetting it by the same raising distance (X) used on all the other rafters. The raising distance is measured along the plumb cut, which is made using the angle shown in blue on the plan for QG. The seat cut for QG is made using the yellow-shaded angle from the plan. Remember that when using this graphic system, the plumb-cut angle is always opposite the run, and the seat-cut angle is always opposite the rise.

Dropping an irregular hip is done as on a regular hip, except that in this case, bisecting the irregular plate angle of this bay means orienting the scrap block at 67½° to the plates, not at 45°. The amount of drop is then measured up from the edge of the plate to the top edge of the scrap block as before.

Next, the unit rafter length from the drawing needs to be stepped off 12 times on the rafter stock to determine the unadjusted rafter length. However, the top plumb cut does not have to be shortened as the regular hip did because the face of header GH, against which QG frames, coincides with its theoretical layout line.

To make the plumb cut at the top end of the rafter, I set a 22½° bevel on my saw (the difference between 90° and 67½°), and cut away from the line. The resulting cheek fits against header GH, and a mirror image of this rafter fits header DG. Spiking these two together formed a V-notch at the top of the double rafter that straddled the 135° angle of the deck corner.

Adding the edge cut—The real challenge of an irregular plan comes when the bevel you are cutting is less than 45°. The bottom cut of valley KR is a good case in point. The bottom end of this rafter frames against the side of hip rafter JR, forming an angle in plan, KRJ, which is sharper than 45°. The solution is to lay out a

Facing page: The author began with the common rafters and the ridges and headers they frame against, and worked up to the more complex parts of the roof, like the dormer and bays (on the left in the top photo), the irregular valley that joins them to the common rafters (in the center of the photo). The bottom photo shows a convergence of two hipped bays, two dormers, and a doubled valley that connects this area to a run of commons.

face cut *and* an edge cut; and then to use either a handsaw or a chainsaw to make the cheek cut through both lines simultaneously (see the sidebar at right). It is in laying out the edge cut that things get a bit rough. I like to understand what I'm doing instead of just using a set of tables on a square, so I turned to Siegele again. His explanation relies on a line he calls the *tangent*, but its not the tangent you learned about in geometry class.

The tangent—This line is used in graphic constructions to find the edge bevel of roof members, roof sheathing, hopper boards, and other compound-angle cuts. Unfortunately, at no point in his books does Siegele give a concise definition of tangent. It's best understood in context, but I've assigned my own definition to the word just to introduce it. Basically, the tangent of a given rafter is an imaginary horizontal line, perpendicular in plan to the run of the rafter, extending from the heel of the rafter to the plane containing the vertical face of that member (ridge, header, or other rafter) against which the cheek cut of the rafter frames. Whew.

As an example, refer to the regular hip rafter LN, seen from above as LD (drawing, below). To

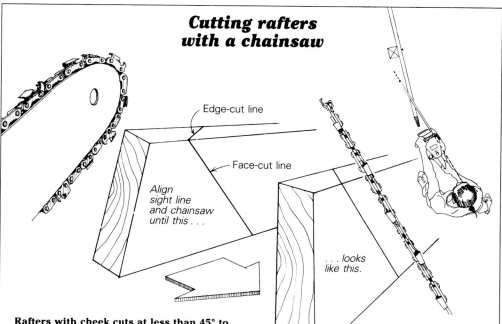

Cutting rafters with a chainsaw

Edge-cut line

Face-cut line

Align sight line and chainsaw until this . . .

. . . looks like this.

Rafters with cheek cuts at less than 45° to their faces must be cut by handsaw or chainsaw. The former is tough work, even on light stock, because you have to cut through the wood at a miter *and* a bevel. This means cutting a cross section greater than the area of the same stock cut square.

To cut rafters with a chainsaw, you must line up the face cut and the edge cut with your eye so they appear as a single line, and then introduce the bar of the chainsaw so that it also lines up along this plane of vision. When everything coincides, you can make the cut.

This all sounds simple enough until you consider that holding the saw as I just described violates the first rule of chainsaw safety: Keep your face and shoulder out of the plane of cut, because that is where the saw will go if it kicks back. But lining up like this is the only way

I know to produce a cheek cut that fits precisely. As a result, I use extra caution when I'm cutting this way. I spike the rafter stock to a post or some other solid object, leaving the end to be cut sticking well out in mid-air. Then I check the stock carefully for nails. I give myself plenty of working room, especially behind me. I keep my arms firmly locked and my legs spread, and drop the saw smoothly through the cut, without twisting. The chain must be kept sharp, well lubricated, and fairly tight on the bar; ripping chain works best.

After sawing, you'll probably want to smooth out the cut you just made with a plane. I use the 6-in. wide Makita 1805B and get excellent results.   —S. M.

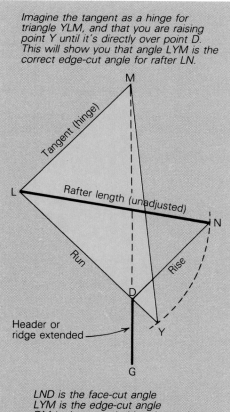

Imagine the tangent as a hinge for triangle YLM, and that you are raising point Y until it's directly over point D. This will show you that angle LYM is the correct edge-cut angle for rafter LN.

M

Tangent (hinge)

L

Rafter length (unadjusted)

N

Run

Rise

D

Y

Header or ridge extended

G

LND is the face-cut angle
LYM is the edge-cut angle
DM is the extended header or ridge

construct the tangent for this hip (LM), I have to draw a line perpendicular to the run of the hip LD that emanates from the heel of rafter LN. The tangent ends at point M, where it intersects the face of extended header DG. This imaginary plane is seen from above as a dotted line—DM—which extends from line DG.

To find the angle for the edge cut, I set the steel point of my compass at L and the pencil point at N and swing an arc down to intersect the extended line LD at Y. Now the extended run LY equals the rafter length. By connecting M to Y, I make angle MYL, which is the edge cut for the hip where it frames against header DG.

Why it works—I'll attempt to explain why this geometric construction works. Imagine that tangent LM, magnified now to full scale, is a sort of giant hinge. Slowly, triangle LYM (colored green in the drawing at left) starts to rise upward, led by point Y, while line LM remains anchored where it is. Your perspective on this is looking straight down.

When line MY coincides in your vision with the plane represented by line DM, and point Y coincides exactly with point D, the plane LYM is

inclined at the right pitch for the hip LN. Only at this pitch is the rafter length LY seen as equal in length to the run LD.

With plane LYM in this angled position, imagine a T-bevel applied with its stock along LY and its blade along YM. Now imagine a hip rafter, with its seat cut already made, raised up into position so that its top edge runs right along the raised-up line LY. If you survived these visualizing gymnastics, you'll begin to see why the angle LYM on the T-bevel gives the correct edge cut on the hip.

Here is another way of describing this concept: Construct a right triangle with legs equal to the tangent and rafter length—the angle opposite the tangent gives the edge cut.

For regular hips and valleys, the tangent is equal to the run. So by substituting the word run for the word tangent, we get a useful corollary: In all regular roof framing, use the unit run of the rafter on one arm of the square, and the unit length of the rafter on the other—the latter gives the edge cut. □

Scott McBride is a carpenter and contractor in Irvington-on-Hudson, N. Y.

Framing an Open-Plan Saltbox

Structural stability can be a problem
when the load-bearing partition is removed

by Pat and Patsy Hennin

Many houses built with energy efficiency and contemporary design in mind just don't look very homelike to lots of people, so builders are taking a second look at the ability of traditional styles to accommodate alternative technology and a modern, open interior while maintaining the dignity befitting a house. The saltbox is inherently efficient, with its high heat-collecting south face and its low north wall, often backed by wind-buffering closets. In some modern incarnations, though, it is also inherently unstable, with structural characteristics that place dangerous outward thrust on bearing walls. Appropriate framing can solve this problem, and result in the best of both worlds—open space within, and the traditional saltbox silhouette outside.

Development—The saltbox style evolved from the center-chimney Cape Cod, 1½ stories with a peaked roof, which was both simple and symmetrical (drawing, facing page, top). The side walls of the traditional Cape are built to equal heights and tied to each other with joists at floor and ceiling levels. Rafters are locked into compression at the roof peak, and the ceiling-level joists neutralize the tendency of the roof load to push the walls out. Above the walls the rafters and joists form triangles, the most solid of shapes, while corner braces (more triangles) give the walls rigidity. A Cape is very stable.

In its pure form, a saltbox is a Cape with a shed addition, which often continues the roof line at its original slope. The original rear wall (A in the drawing) becomes a shared load-bearing partition, and the new roof exerts no outward thrust because shed roofs place their loads equally on both bearing walls (A and B). An old-fashioned saltbox is as stable as the Cape it evolved from.

Today's saltboxes, though, are usually erected from scratch by builders who want a traditional look outside, but large, open spaces inside—and this can be the source of many structural problems. The saltbox's exterior walls and rafters by no means form a free-standing arch, and dispensing with the common partition or the joists that tie the rafters together destroys its structural integrity. Under heavy loads, such a building will collapse. If you want a sound, open-plan saltbox, you have to compensate for these missing structural elements.

Framing a new saltbox—One way to deal with the outward thrust of the roof is to eliminate rafters and build the entire structure in bents of progressive heights conforming to the shape of

A modern saltbox under construction in upper New York State. Framing is basically traditional, but the common partititon between Cape and shed has been replaced by posts and beams to open up the interior. Photo: Patsy Hennin.

the saltbox roofline (drawing, facing page, bottom). No rafter ties are necessary, so the house can be open vertically as well as horizontally—great if you want a cathedral ceiling. Each bent is made up of a purlin (a timber running parallel to the ridge) resting on posts that help form the end walls, which also become the bearing walls. (In a conventional Cape, the front and rear walls are the bearing walls.)

Purlins up to 18 ft. long are cost-effective, so assuming the tall wall is to the south, this method can provide clear spans from east to west of up to 18 ft. You can add sections of up to 18 ft. to the east or west. If you have special needs, you can replace one or two common posts with a beam that will support the weight of purlins. If you need to span more than about 12 ft. under loading conditions similar to Maine's, look into using a steel beam. Steel companies will size them for you, and they probably won't ruin your budget.

There are probably as many ways to frame up a new saltbox as there are housebuilders, but let's run quickly through one way you could proceed, taking as our example a house that is, in effect, two saltboxes of different heights slightly offset. This is probably the hardest case, and if you can frame it, you'll be able to frame any similar design.

The foundation must support the weight of the loaded building in the soil of the site and against heavy wind loads. For example, we can use pressure-treated 6-in. to 9-in. diameter utility poles anchored to concrete pads below frost level, attaching heavy 6x10 sills to them with steel strapping, as shown in detail C. In this case,

we would space the posts 8 ft. o.c. from north to south and 9 ft. o.c. from east to west, so that a row of them would run front to back under the center of each part of the structure. You could, of course, pour a perimeter wall or a slab, and make the sill connection with bolts set right into the concrete.

Joist hangers are the simplest way to connect joists to sills, but they are expensive and aren't sized to accept roughsawn lumber. We usually set the joists 16 in. o.c. on top of the sills. We use 16d nails to attach 2x10 headers to the joists' ends, take careful measurements corner to corner, and use a sledge to persuade the whole shebang into square. Then we toenail the joists into the sills with 10d nails. Subflooring (plywood, diagonal boards or flakeboard) is fastened to the joists with 8d nails, and we've got a platform to work on.

The post-and-purlin bents can go up as units, but putting up the side walls first makes it easier to keep the roofline straight. The walls are framed right on the deck. We usually use 4x6s, 4 ft. o.c., with 2x6 top and bottom plates and install horizontal blocking every 4 ft. up the height of the walls to act as a nailing surface for asphalt-impregnated sheathing, which breathes better, prevents infiltration better and is cheaper than plywood. Bracing against wind racking can be achieved with either plywood in the corners or diagonal steel or wood bracing. If you are framing with roughsawn lumber, 20d galvanized box nails are standard fasteners. Use 16d common nails with milled wood. Tilt the side walls up, nail the bottom plates through the deck into the joists with 20d nails, and brace the walls with 2x4s or the like while you get the 4x10 purlins up and in place.

Small diagonal blocks under each purlin will keep them all sitting perpendicular to the floor, the attitude in which they're strongest. One of the best ways to secure the purlins is to use a ⅜-in. self-feeding 18-in. carbide bit to drill two holes through purlin, block, top plate, and 6 in. into each post. Use a sledge to drive ⅜-in. rebar into each hole to anchor the purlins solidly in place (detail B). Toenail 2x10 blocking to the top plate between purlins, trimming it to the proper height. This blocking helps keep the purlins from twisting. Extending the purlins 18 in. or so beyond the posts allows for overhangs at east and west.

Where the purlins of the lower house section lie against rather than on top of posts, we notch the posts out 2 in. deep (detail A), and also bolt

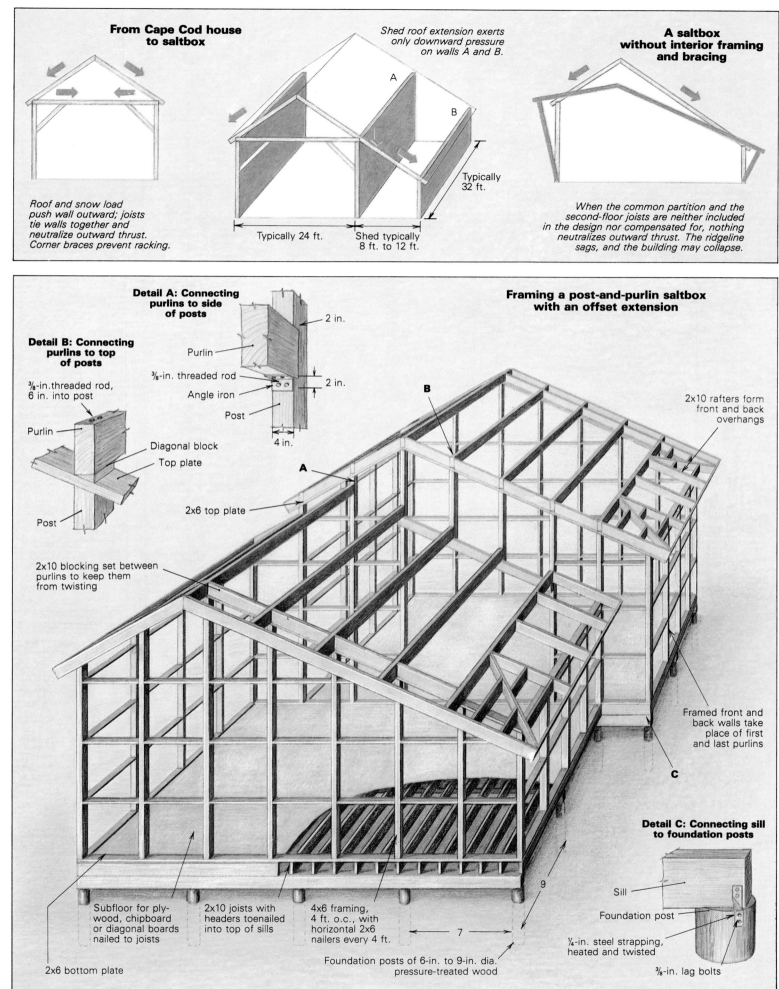

Illustrations: Lee Hov

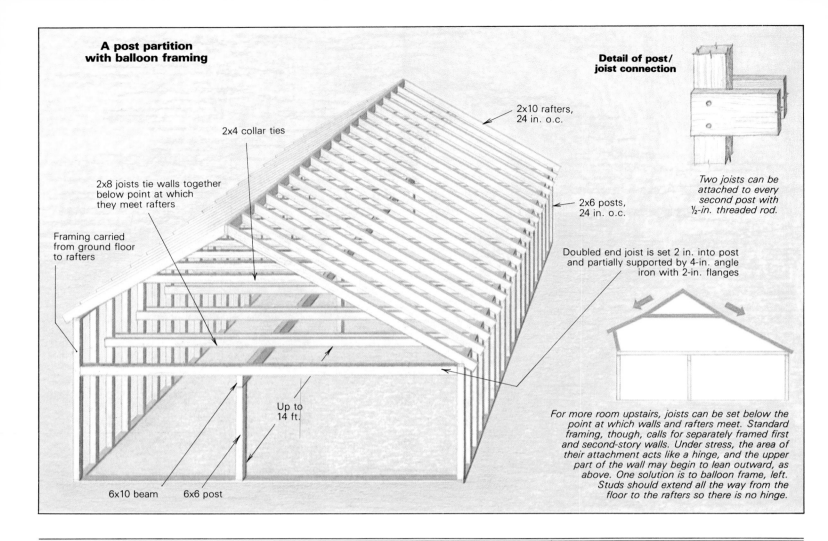

A post partition with balloon framing

2x4 collar ties

2x8 joists tie walls together below point at which they meet rafters

Framing carried from ground floor to rafters

6x10 beam   6x6 post

Up to 14 ft.

2x10 rafters, 24 in. o.c.

2x6 posts, 24 in. o.c.

Doubled end joist is set 2 in. into post and partially supported by 4-in. angle iron with 2-in. flanges

Detail of post/joist connection

Two joists can be attached to every second post with ½-in. threaded rod.

For more room upstairs, joists can be set below the point at which walls and rafters meet. Standard framing, though, calls for separately framed first and second-story walls. Under stress, the area of their attachment acts like a hinge, and the upper part of the wall may begin to lean outward, as above. One solution is to balloon frame, left. Studs should extend all the way from the floor to the rafters so there is no hinge.

Sizing a beam

A beam must be large enough to support the "dead" load of the structure above it and the "live" load (of snow or activity) that is imposed on it. Calculating the size of a beam is a two-step process: First, add up the various loads on the structure and decide how they are bearing on the beams. The weight is either spread evenly (a uniform load) or concentrated (a point load). A point load typically requires a beam twice as large as the same uniform load would. Second, choose a beam of the right shape and species to resist the load.

The first load calculation is a bending moment problem in engineering terms. More than 40 formulae analyze the ways a load tends to break a beam, depending on the type of load (uniform or point), and on the beam and how it is supported (at both ends, cantilevered, with ends overhanging and so on). *Architectural Graphic Standards* is a useful reference for typical beam placement formulae; your local library probably has a copy.

As an example, let's size a purlin like one shown in the drawing on the previous page. This purlin is supported at both ends, with a roof load and a snow load spread evenly along its clear span (unsupported length)—a simple beam with a uniform load. The bending moment (M), or load that will break it, is calculated by:

$$M = \frac{WL}{8},$$

where W is the weight carried by the beam (the load per square foot multiplied by the top surface area of the beam in square feet) and L is the unsupported length of the beam in inches. Assuming that the posts are spaced 4 ft. on center, this beam will carry a rectangle of roof area above it 4 ft. (halfway to its neighboring purlins) by 18 ft. (the length of the clear span), or 72 sq. ft. On each square foot is the weight of the snow plus the weight of the roof structure itself. Check your local building code's snow-load estimate; in Maine it is 40 lb./sq. ft. The weight of the structure varies, but a typical figure is 10 lb./sq. ft. Hence the total weight on this purlin is 50 lb./sq. ft. Returning to the formula,

$$W = 50 \text{ lb./ft.}^2 \times 72 \text{ ft.}^2 = 3,600 \text{ lb.}$$

$$L = 18 \text{ ft.} \times 12 \text{ in./ft.} = 216 \text{ in.}$$

$$M = \frac{WL}{8} = \frac{3,600 \text{ lb.} \times 216 \text{ in.}}{8} = 97,200 \text{ in.-lb.}$$

The amount of wood needed to resist this load is the section modulus (S), found by dividing the bending moment M by the fiber stress value (f) on the species of wood, or the pounds per square inch that the species can carry. *Architectural Graphic Standards* and many wood and building guidebooks list allowable stress values for structural timber of commonly used species. Eastern hemlock, our usual choice, can carry 1,200 lb./sq. in. This section modulus is calculated as follows:

$$S = \frac{M}{f} = \frac{97,200 \text{ in.-lb.}}{1,200 \text{ lb./in.}^2} = 81 \text{ in.}^3.$$

The section modulus is just a theoretical shape. The actual concrete dimensions of the lumber are calculated by the formula

$$S = \frac{bd^2}{6},$$

where b is the breadth in inches and d the depth in inches of the beam. We usually use roughsawn lumber, where a 2x4 is really 2 in. by 4 in.; for milled lumber, use the planed-down dimensions. To find the right size beam, substitute various sizes for b and solve for d. The strength of the concrete dimensions must be equal to or greater than the theoretical shape, in this case 81 in.³. For example, if b is 4 in., d is 11 in. If b is 6 in., d is 9 in. Since lumber is sawn in 2-in. increments, order a 4x12 or a 6x10. Either will do. There are tables that will size joists and beams, but rarely can you find one that will size a timber as large as these purlins. Also, most tables aren't written for roughsawn lumber. —*Patsy Hennin*

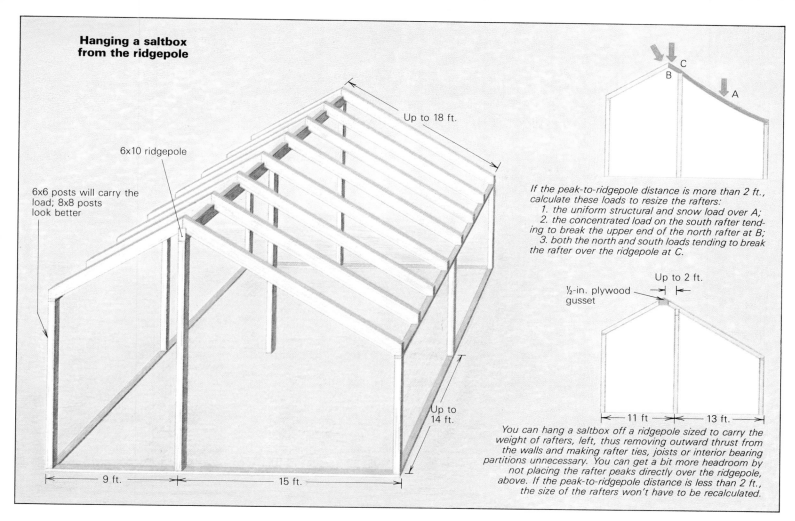

Hanging a saltbox from the ridgepole

6x10 ridgepole

6x6 posts will carry the load; 8x8 posts look better

Up to 18 ft.

Up to 14 ft.

9 ft.

15 ft.

If the peak-to-ridgepole distance is more than 2 ft., calculate these loads to resize the rafters:
1. the uniform structural and snow load over A;
2. the concentrated load on the south rafter tending to break the upper end of the north rafter at B;
3. both the north and south loads tending to break the rafter over the ridgepole at C.

½-in. plywood gusset

Up to 2 ft.

11 ft.

13 ft.

You can hang a saltbox off a ridgepole sized to carry the weight of rafters, left, thus removing outward thrust from the walls and making rafter ties, joists or interior bearing partitions unnecessary. You can get a bit more headroom by not placing the rafter peaks directly over the ridgepole, above. If the peak-to-ridgepole distance is less than 2 ft., the size of the rafters won't have to be recalculated.

angle iron with 2-in. flanges onto the posts with ⅜-in. threaded rod at the bottom of the notch, creating the 4-in. bearing surface required by most building codes.

We don't set purlins on top of the first and last posts. Instead, we frame the 4x6 front and rear walls up to the proper height. We set 4x6s 4 ft. o.c. with 2x6 top and bottom plates and blocking, just as in the side walls. The asphalt-impregnated sheathing goes on with 1½-in. roofing nails. We also extend rafters from the front and back purlins to create overhangs at the north and the south.

On the roof, 1-in. boards over the purlins do double duty as both sheathing and ceiling. They are nailed to each purlin with three 8d nails. This ties all the purlins together and further reduces the likelihood of their twisting. At the peak, we fasten boards from both sides of the roof together with 1⁄16-in. or 22-ga. metal secured with 5d nails. Our post-and-purlin saltbox is now framed and sheathed.

Modifications to traditional framing—There are techniques other than post and purlin that can result in a structurally stable open-plan saltbox. One is simply to modify traditional framing by replacing the common wall with a series of structural posts topped by a beam sized to support the rafters above (drawing, facing page). Here in Maine, we can use reasonably sized beams to span up to 14 ft. In areas where a smaller snow load can be expected, you could span greater distances. This technique results in

an open horizontal space, but the joists tying the walls together block vertical openness.

If you want an open first floor and also more room on the second, set the joists below the level at which the rafters meet the top plate. Then install collar ties near the peak of the roof. You'll still need the common partition or a row of posts, but this technique increases the headroom at the south wall, and will give you more space for upstairs rooms. It imposes great point loads on the south studs, though, and snow loads could make the south wall bend at the plate between the first and second floor framing. The best solution is to balloon frame—use studs long enough to carry in one piece all the way from the first floor through the second floor to the rafters, and use enough of them to divide the loads down to a minimum per stud and rafter. In Maine we often use 2x6s, 24 in. o.c. as studs, and set our rafters 24 in. o.c. also.

Hanging from the ridgepole—Another way to eliminate outward thrust is by using a ridgepole sized as a beam to carry the weight of the rafters (drawing, above). Sizing a beam is explained in the box on the facing page. If you attach the rafters to the ridgepole with angle iron or rebars, or gusset them to each other over the ridgepole, the building will hang like a tent from the ridge. You no longer need second-floor joists or rafter ties, because there is no outward thrust. As with the purlin method, the house can be open vertically as well as horizontally. Rafters up to 18 ft. long make economic sense, and in

Maine ridgepoles can span up to 14 ft., so clear interior spans in a saltbox could reach 24 ft. by 14 ft., and the only obstruction in a 24-ft. by 28-ft. room would be a single post in the middle.

If you need more space north or south of the obstruction caused by the rafters and ridgepole, you can set the peak several feet away from the ridge, as in the drawing, above right. The southern rafters are in compression against the northern rafters, and are exerting a downward pull on their high ends. Technically, this is a point load on cantilever, and the ridgepole can be as much as 2 ft. from the peak without affecting the size of the rafter. If you want the peak even farther from the ridge, you have to consider three types of load when you calculate the size of the rafter, as shown in the drawing, top right.

You can build a safe, open-plan saltbox using any of these methods, or a combination of several. Draw the floor plan first and think about which technique would obstruct your pattern least. If necessary, fiddle with your floor plan a little so that you can use a method that will guarantee a stable house. If changing your floor plan is too traumatic, consider using steel beams to allow longer safe spans. Remember that you always have to account for the outward thrusts created by disturbing the original simplicity and balance of the Cape/shed combination that we now call the saltbox.   □

The Hennins are directors of the Shelter Institute in Bath, Maine, an owner-builder school which they founded in 1973.

Trouble Spots in 19th-Century Framing

Old-time workmanship wasn't always what it's cracked up to be. Here's what to watch for if you're rehabilitating an old house

by Dan Desmond

Too many of us involved with preservation work are caught up in a nostalgic reverence for that which never was. Most people who thump the walls of an old house and lament that "they don't build 'em like they used to" have a vision of the old housewright, disciplined master of his craft, who did it right and built it to last. If those walls belong to a 17th or 18th-century house, such confidence is probably warranted, but if the house was built in the 19th century, it may be purest speculation. To some people longevity is admirable in itself. I love old houses, too. But as one who lifts them in the air and has lived to tell the tale, I find occasion to criticize their structures. I manage a project for the Keene (New Hampshire) Housing Authority, which is funded by the Department of Housing and Urban Development to provide public housing through what is appropriately called Substantial Rehabilitation. We work on old houses that have been shunned by the private market, houses that are depressing values in otherwise stable neighborhoods. These are "worst case" examples of neglect—vandalized, dilapidated and

condemned. By rehabilitating them, we provide public housing, eliminate eyesores and demonstrate that few houses are really so far gone that they should be demolished.

In our work, we have uncovered a virtual casebook of 19th-century framing, from about 1830 on. Some of our discoveries are peculiar to New England, but 19th-century framing practice reflects social, economic and technological imperatives that were national in their effect.

The roots of the problem—While many houses in the last century were built with skill and care, many more were not. To understand how this could happen, considering the tradition of craftsmanship that had gone before, remember that 19th-century America was enjoying an unparalleled rate of new commercial and industrial expansion. Timber-frame construction was too slow and expensive for the market, which was gobbling up dimensional lumber as fast as the new steam and water-powered mills could supply it. The Industrial Revolution, with its emphasis on uniformity, high volume and low cost,

made owning a house a new and tantalizing option for the common man. (The first half of the 19th century saw a 300% reduction in the cost of lumber, and an 800% drop in the cost of nails.) Naturally, a lot of people wanted to build houses, including some enterprising do-it-yourselfers and others who were a bit short on experience. And, like many house builders today, they put their money into eye-catchers, paying less than close attention to things buried behind the walls.

During the 1800s, the plasterers and finish joiners were the great craft artists. Their compensatory skills at shimming and feathering covered a multitude of framing irregularities. Exterior appearance and finish work notwithstanding, you may uncover a house that was thrown together and has held together for all the wrong reasons. Accepted standards for framing practice and load bearing were not widely acknowledged by carpenters until late in the century. Even then, old-timers were not always at ease with the balloon frame, and retained random elements of timber-frame design in the houses they built. That's why you will seldom find a

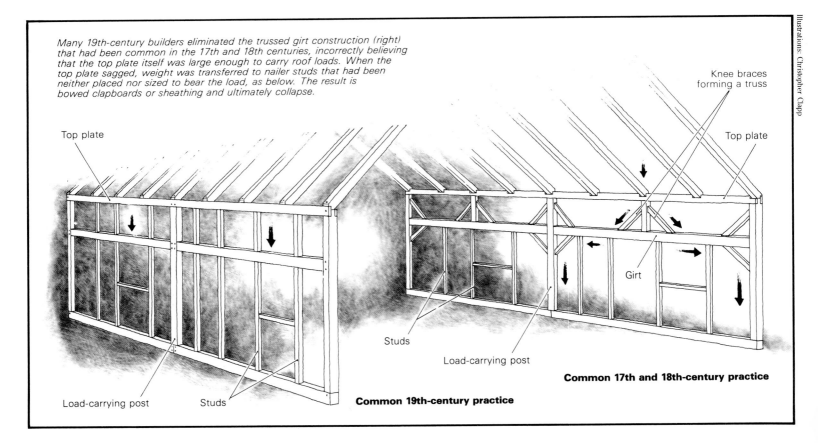

Many 19th-century builders eliminated the trussed girt construction (right) that had been common in the 17th and 18th centuries, incorrectly believing that the top plate itself was large enough to carry roof loads. When the top plate sagged, weight was transferred to nailer studs that had been neither placed nor sized to bear the load, as below. The result is bowed clapboards or sheathing and ultimately collapse.

Top plate

Knee braces forming a truss

Top plate

Studs

Girt

Load-carrying post

Load-carrying post   Studs

Common 19th-century practice

Common 17th and 18th-century practice

Illustrations: Christopher Clapp

From *Fine Homebuilding* magazine (December 1981) 6:36-38

pure example of any one framing system like the ones you see in the fix-it books. You can and will find anything behind the walls, and in repairing or rehabilitation, you will have to rely on your skills, not those of someone who might have been in a hurry 150 years ago. You don't want to dig into a vital and usually expensive job and find that what appeared to be a single weakness is inherent in the frame.

Checking the walls—Before beginning any structural repair to a house built 100 or 150 years ago, you've got to know the condition and the composition of its walls. This is especially important if you're facing a repair to some portion of the foundation—the most common job required on a house of this age—because safe and effective lifting and supporting calls for a thorough knowledge of the condition of the frame. This goes far beyond the cursory beam-stabbing ritual practiced by most termite inspectors. An 8x8 sill can be rotted from the outside a good 6 in. and still look fine from the cellar.

For a thorough exterior inspection, carefully examine the sill, and pay particular attention to corner joints. If the house is timber-framed, remove the clapboards and subsheathing from the girt between the first and second floor, both to make sure that the structure is sound and to check that the wall studs or posts are continuous up to the top plate.

The absence of continuous, aligned uprights is not at all uncommon in 19th-century houses. Studs were merely nailers and did not usually bear heavy loads in well-built 17th and 18th-century houses. Their placement wasn't critical. A trussed girt at the top plate carried loads to the large posts that were sized to take them (drawing, facing page). But many 19th-century builders eliminated the time-consuming truss assembly in the belief that the massive top plate would still carry roof loads. As it sagged, this plate put enormous strain on studs that weren't meant to withstand it. Carpenters got away with

this because the clapboard, subsheathing and lath acted as the sides of a box beam, and the arrangement was often strong enough to hold things in place even when the foundation collapsed. Ultimately, of course, such a structure can fail. One easily recognizable sign of this is a conspicuous bowing of clapboards or outside sheathing. Houses have survived for decades in this condition, but might not survive major sill or foundation work unless their frames are first repaired. These problems are less common in late 19th-century houses, after the techniques of balloon framing had become refined.

Check interior walls—Inside, make no assumptions about which partitions are load-bearing. Examine the juncture of wall and ceiling to determine if there is a beam, or even a top plate to carry and distribute loads (photo below left). Since this usually involves removing some plaster and lath, discretion and care are needed to minimize damage. This technique may frighten you if you are repairing a house of architectural or historical importance, but we do it because we believe that it's better to patch a little plaster than to begin work assuming that a partition wall is load-bearing. This mistake is easy to make because of a number of old-time framing practices that you're not likely to see in modern platform construction.

First is the practice of running the floor and ceiling joists parallel to the ridgepole instead of perpendicular to it, the way we do today. This was probably an effort on the part of framers to distribute the floor loads on the gable end of the house, while concentrating the roof loads on the eave end. In Victorian times, the original wood shingle roofs of houses framed this way were often replaced with slate. This put an extraordinary strain on the system. If you're working on a house like this, you should install collar ties as low as possible between its rafters (drawing, right). Be especially careful when insulating or wiring an attic with this type of construction, be-

cause the floorboards act as the bottom chords of a truss with the rafters, and removing too many of them could spring the sidewalls. I have seen 8x8 top plates split their full length and tie beams broken at the mortise because the insulating contractor saw no reason to renail the floor (photo below right). If you must gain access to the joist cavity in this kind of frame, lift the boards one at a time, renailing as you go. Also, don't assume that because the joists run in a certain direction on one end of the house, that they run the same way on the other. They may alternate, room to room and floor to floor.

The second practice is the persistent use of timber-framing techniques. Despite the increasing popularity of lightweight dimensional lumber, old-school carpenters would set heavy beams in clear span from one side of the house to the other. What we would take for a load-bearing wall is often nothing but slab or waney lumber turned sideways, faced with lath and topped with a thin piece of scrap. Over the years, as the structure settled, these partitions often assumed a role of support that they weren't designed for. In our projects, we remove these

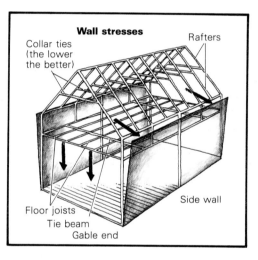

Wall stresses
Collar ties (the lower the better)
Rafters
Floor joists
Tie beam
Gable end
Side wall

In 19th-century houses, walls assumed to be load-bearing often are not. Underneath the plaster and lath, left, boards are simply nailed to the bottom of a light scrap timber. The wall was never meant to support weight. In the old house at right, the attic floor joists run parallel to the ridge. When the floorboards were removed to install insulation, the side walls sprung, splitting the great top plate along its length and breaking the joint between it and the tie beam. The drawing above shows the forces at work, and also the collar ties that would help to hold the system together.

Some 19th-century practices call for 20th-century repairs. Top left, windows and doors cut into a house's heavy sheathing are nailed into place with stud cleats (in this example, the jack studs don't even reach the window frame). In the absence of headers, stresses are transferred to the house's skin, not to its frame. Lifting the house for foundation repairs without bracing such openings can be disastrous. Right, floor joists are face-nailed together. Floorboards (or siding, in the case of spliced wall studs) can hold such joints together, but when this reinforcement is removed during rehabilitation, the frame may fail. In the 19th century, builders sometimes salvaged beams from older structures. Such timbers have often been weakened by mortise holes, and if they are in span, like the one above left, they have to be supported if they are to be left in place.

walls and start with new supports and partitions from the ground up. Where preservation is the object, you could install a small-diameter lally column (a steel tube, filled with concrete and cut to length) in the wall to support the sagging beam, providing that the column itself is supported under the floor.

More problems—Two other framing practices were common in the 19th century, and you should always look for them, especially before doing any lifting. First is the failure to use window or door headers in many old houses. Sometimes large wall areas were spanned with a single beam, and windows and doors were cut into the heavy sheathing and nailed into place with stud cleats that often didn't contact the sill and girt (photo top left). This can give you fits if you have to replace, say, the sill in a house with a lot of windows, because during a lift the loads transfer to the lath and clapboards in an unpredictable fashion. If possible, brace doors and windows without headers, remove window sashes to avoid their racking, and concentrate lift points between windows.

The other is the practice of splicing. You will

see this most often in early attempts at balloon framing. Carpenters would fit into place the longest stud that they had. If it didn't reach the plate, they would tack a piece of scrap on the side to complete the support. As I've already mentioned, the tension of the siding or clapboards may be holding a house like this together. If you remove them during a thorough rehabilitation without reinforcing or replacing the splice, the frame might well fail.

The same technique was used on floor joists on occasion (photo above right). Since nails aren't very strong in shear, you have to be careful not to concentrate any new loads on these members without reinforcing the splices.

Another behind-the-wall surprise is the practice of "filling in." In colder areas of the country builders would often fill the outside wall cavity with lath and rough plaster, or occasionally solid brick, to reduce winter air infiltration. This can cause tremendous problems today when you want to wire or insulate the area. Brick-filled walls are especially troublesome if your repairs involve lifting or leveling.

You will also find houses in which most structural members appear to match, and then see a

used timber from a far older structure pieced in. This tells you something about old-time builders: First, they were practical and frugal. Second, they were not nearly as much in love with the practice of hand-hewing as some would have us believe, and were only too happy to get some extended use out of Colonial-era salvage. These members usually worked well as sills or posts, and you can probably leave them in place. In span, though, they often failed because they had been weakened by numerous deep and unfilled mortises. Such timbers should be replaced or reinforced (photo above left).

Economic conditions and a new appreciation for our past will make the purchase and rehabilitation of old houses a growth industry in the 1980s. The necessity for structural repair will expose many new owners to the flip side of old-house charm. Such work is hard, and it's never successful unless it's also safe. That requires a realistic and critical approach to the qualities and styles of workmanship behind the walls. □

Dan Desmond lives in an 1840 timber-framed house in Walpole, N.H. He has been a builder and renovator for 15 years.

A Shed-Dormer Addition
Extensive remodeling triples the space of a small Cape

by Bob Syvanen

Adding new space to an old house is seldom as simple as it first looks. Getting the end result to look good is demanding and time-consuming, and the problems that invariably crop up add considerably to the headaches. Working out the basic design and particular construction details on paper isn't so hard, but once the actual work begins, leveling and plumbing the new construction to fit the old structure is a big challenge, especially if years of settling have caused walls to lean and beams to sag. Apart from all these difficulties, keeping the house and site relatively clean and orderly during construction so that life can go on for the clients who are putting up with this invasion adds appreciably to the burden and the time spent.

I faced such a challenge on a recent job. My clients asked me to take their plain 864-sq. ft.

Cape Cod cracker box and make it over into a not-so-plain 3,300-sq. ft. house. Though I had considerable freedom with the design, the clients made several specific requests. Chiefly, they wanted about 1,300 sq. ft. of the addition to be on the second floor. There they would set up a large master bedroom, something they had missed since moving to the East Coast from California. Since the house would be occupied during construction, I had to keep the old Cape tight to the weather during the entire process of adding on.

I decided to keep the front roof intact and to build the addition at the back and south side of the old house. The floor plan I worked up (drawing, next page) enlarged the old kitchen and added a dining area, a living room, a bedroom, a bath and a foyer—all on the first floor. I added a three-car attached ga-

rage on the north end of the house. Upstairs went the big bedroom (about 500 sq. ft.) along with a bath and access to these via a new stairway. The other part of the upstairs (the space directly above the old house) remains unfinished even now, but will eventually become another bath and bedroom.

The owners wanted to keep these two areas entirely separate, thinking that the unfinished area might become an in-law apartment in the future. To this end, each space on the second

While a two-story addition to this small house was being built, the old roof had to be kept intact until the addition was complete. The 2x10 nailed along the ridge of the old roof serves as a plate for the rafters that extend to the new ridge. Kicker braces for plumbing the dormer wall are still in place.

From *Fine Homebuilding* magazine (December 1984) 24:47-51

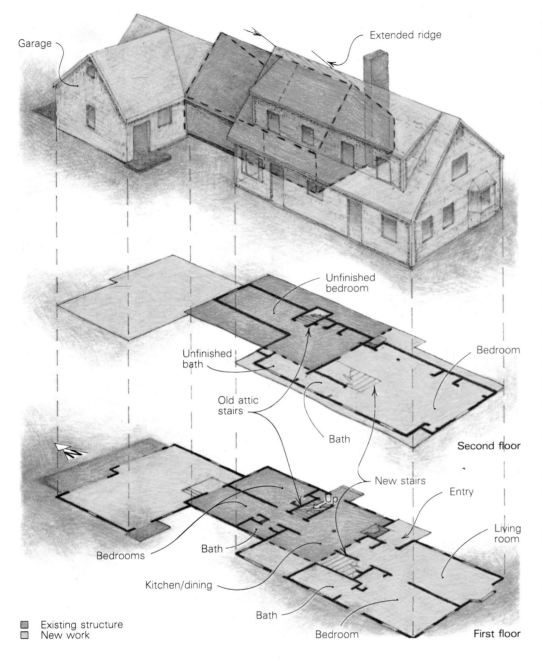

Plan of old house and additions

Garage

Extended ridge

Unfinished bedroom

Unfinished bath

Old attic stairs

Bath

Bedroom

Second floor

New stairs

Entry

Up

Living room

Bedrooms

Bath

Kitchen/dining

Bath

Bedroom

First floor

■ Existing structure
☐ New work

---

story has its own way up and down. Excluding the garage addition, remodeling the old house extended the south gable-end wall 30 ft. and added an 11-ft. wide by 18-ft. long section in the opposite direction, along the existing back wall behind the kitchen.

By putting a 40-ft. long shed dormer on the rear of the house, I was able to create the large, open second-story space the clients wanted. Doing this and making the house wider from front to back meant extending the plane of the roof on the front of the house and relocating the ridge so that it would exactly divide the new footprint along its width. The challenge I faced in framing the new roof was to build the dormer big enough for the owners' wants, strong enough for the open floor plan (the ceiling joists had to span almost 20 ft.) and make it look good from the outside. And, of course, I had to do it all without opening the roof to the weather.

Maybe the most trying part of the job was

an uninvited but regular member of the crew named Jasper, a kleptomaniac crow. He seemed to like nail sets best and would pick them right out of your tool belt. We must have lost a dozen nail sets. We even had to keep the truck windows rolled up to keep him from flying off with our ignition keys.

The old attached garage had to be removed from the south-end wall to make way for the addition. But before doing that I had to build the new garage at the opposite end of the old house. Tearing away the garage left a wound in the gable end open to the weather. It also exposed the mudsill, from which I would calculate the height of my new floor. The old and new floors would have to match in height. The old mudsill is a 3x6, the floor joists are 2x8s, the subfloor is ⅝-in. plywood, and the finished floor is ¾-in. hardwood, totaling 11⅜ in. For the addition, the mudsill is a 2x6, the floor joists are 2x10s, the subfloor is ⅝-in. plywood, and there is ½-in. underlayment for

carpeting, totaling 12⅛ in. All floors would be carpeted, so the ½-in. underlayment in the addition would need to be level with the finished floor in the old house. Adding ½ in. for grout under the mudsill, I knew that the top of the foundation for the addition would have to be 1¼ in. below the old foundation wall.

I like to level and seal my mudsills with a bed of mortar between sill and foundation. It's a big help on remodel work to have this half-inch to play with, and can often make the difference between being able to adjust wall height properly and having to live with roller-coaster top plates.

When I told the foundation contractor to set the new foundation parallel with the front wall of the old house, which is also parallel to the ridge, I was thinking ahead to tying the new roof into the old roof, wanting to avoid any torqued planes and a snaky new ridge. Apparently he misunderstood my instructions (or just ignored them), and instead made the gable-end wall of the new foundation parallel with the gable-end wall of the old house. He then squared up his forms so that the new sidewall foundation was perpendicular to its end wall. Just exactly the opposite of what I needed. The garage gable-end had been built completely out of square with its sidewalls. As a result, the foundations for the new sidewalls were so far out of parallel with the sidewalls of the old house that Jasper the crow might have been able to do a better job.

To compensate for the misalignment without calling for a jackhammer and a new pour, I offset the mudsills on the new foundations. But after cantilevering the sill as far off the foundation wall as I dared, I found that it was still about an inch away from being parallel with the front wall of the old house. So I decided that I would just have to tolerate the error, and cope with the consequences when I framed the new roof.

With this problem postponed, I completed the first floor of the addition with little hassle. I took care to level my mudsills, to square the new floor before nailing down the deck, and to plumb the corners of the exterior walls to maintain alignment with the old house.

The next thing I had to worry about was the framing for the 11-ft. by 18-ft. area that would enlarge the kitchen along the back wall of the house. After the addition was fully enclosed, this section of wall would be taken out, but for the time being I was stuck with it. The ceiling joists in this area run from the new back wall to the old back wall of the house. To carry them, I nailed a doubled header to the top plates along the old back wall. To make room for the header (or carrying beam), a 2-ft. wide section of the eave had to be removed and the rafters cut back (photo facing page, left). This left a gap in the roof open to the weather. To cover the gap, I tucked an 18-ft. long piece of building paper under the roof shingles and lapped it over the header. A 20-ft. length of old gutter, hung below the beam, channeled the water run off from the old roof into a length of PVC downspout angled out a window opening of the addition.

New joists, new rafters. The addition's second-story floor joists that tie into the old house (above) are attached to a header nailed to the top plates of the existing wall. To make room for the header, the eaves and rafters had to be cut back. Extending the plane of the old roof without opening it up meant framing an 18-ft. long section of roof with a bottom plate on the deck (right), as though it were a wall. Then the section was raised and braced in position alongside the old ridge. The bottom plate was nailed to the old roof at the ridge, with allowance made to let the new roofing work out flush with the old. A layout error in the foundation was compensated for at the peak by sistering a 2x10 onto the new ridge so it would line up with the already-installed gable rafters.

Framing the dormer—With the first-floor ceiling joists and plywood deck in place, I was ready to tackle the dormer framing. I took measurements of the old roof rafters and made a full-size layout on the deck. This layout included the old rafters (many of which would be removed after the framing was finished), new rafters, first and second-floor ceiling joists, the dormer window wall and the interior wall that supports the new ceiling and the old roof. To complete the layout, I added the fascia and soffit. From this layout, I would get the correct angles for making plumb and level cuts on my rafter stock, and the exact lengths for rafters, joists and studs. The full-size layout also let me preview the operations to come and gave me a good sense of how things would go together.

Using the stud height that I had laid out on the deck, I built the 40-ft. dormer window wall with a 6-ft. wide piece of the cheek or end wall at each end (drawing, facing page). This partial return on each side let me brace the new dormer wall at both ends so that it was ready for rafters, and still take my time in filling in the rest of the cheek wall. I set the dormer window wall back 4 ft. from the exterior wall below, and set the dormer's cheeks in 4 ft. from the gable-end walls. To carry the tops of the short rafters, I nailed a 2x8 ledger board to the studs of the window wall 30 in. up from the bottom plate. I think dormer walls look better when they are set back from the edge of the roof. It's more costly to build a dormer this way, but it maintains the integrity of the original roof line, and incorporates the dormer better into the rest of the structure, making it seem less like an ungainly afterthought.

I decided to preserve the plane of the original front roof, and extend it to a new ridge about 4 ft. higher than the existing one (photo

above right). This was a practical measure that would let me take advantage of the existing framing without exposing the old house to the elements. The trick here was being able to extend this plane up to the new ridge without sistering new rafters onto the old ones, which would mean cutting into the roof and removing the old ridge.

To accomplish this, I framed a section of roof to come off the old ridge at the original pitch and in the same plane as the front roof. I began by cutting the 7-ft. long rafters from 2x8 stock, taking the lengths and angles from the layout. Next I nailed a 2x10 plate to the bottoms of rafters (which were set out on the deck), just as if I were framing a stud wall with an oddly angled bottom plate. Then I nailed the 2x12 ridge to the rafters. When this was done, I raised the whole thing into position, and braced it with 2x6s from the dormer deck.

Once the rafter section was in place, I nailed the plate to the existing roof next to the old ridge, as shown in the drawing next page, top left. Because the rafters underneath the plate would eventually be cut off, I made sure that I got plenty of nails through the plate and into the ridge itself; otherwise, I'd have to depend on the nailing in the old rafter stubs to hold the new roof sturdily in place. I was careful to locate the plate so that its top edge was in line with the top edges of the rafters on the opposite side. This way the new roof deck would flush up with the existing deck, and there wouldn't be a hump in the shingles.

To brace the unsupported rafters, I nailed a straight 2x6 guide on edge to the old roof at each end of the new rafter section and let the ends run to the new ridge. By deducting the thickness of the decking and the thickness of the shingles on the old roof, and by cutting shims to this thickness and nailing them to the guides, I was able to find the right plane for the new rafters to lie in. Since the old roof would have to be reshingled, I didn't worry

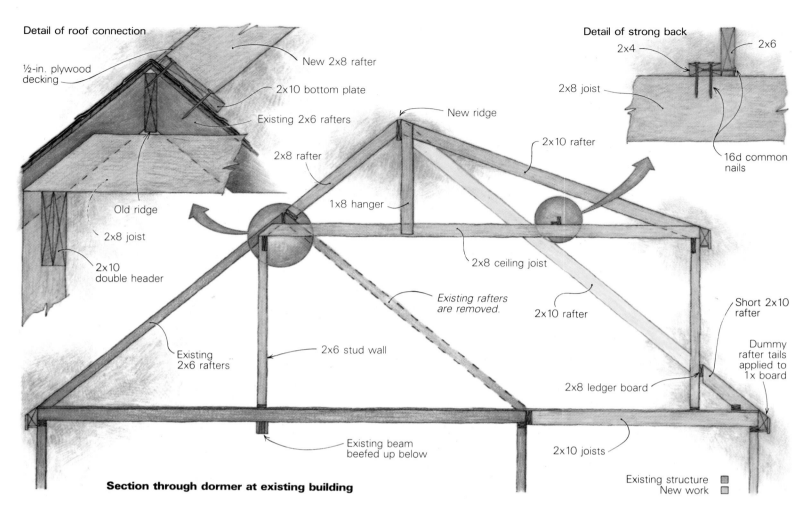

Detail of roof connection

½-in. plywood decking

New 2x8 rafter

2x10 bottom plate

Existing 2x6 rafters

Old ridge

2x8 joist

2x10 double header

Existing 2x6 rafters

Detail of strong back

2x4

2x6

2x8 joist

16d common nails

New ridge

2x8 rafter

1x8 hanger

2x8 ceiling joist

Existing rafters are removed.

2x10 rafter

2x10 rafter

2x6 stud wall

Short 2x10 rafter

Dummy rafter tails applied to 1x board

2x8 ledger board

Existing beam beefed up below

2x10 joists

Existing structure ▦
New work ☐

Section through dormer at existing building

much about nailing the temporary guides through the roofing.

Before putting the shed-dormer rafters on, I wanted to be sure the ridge on the extended rafter section lined up with the new gable-end rafters. So I cut a pair of gable rafters, and temporarily set them in place using a 2x4 spacer in place of the ridge. Once again, I got the gable-rafter lengths and angles from my layout on the deck. Then came the moment of truth, the moment I'd prepared for with all the fussing, fudging and plumbing up the frame from the mudsills below. Eyeballing down the 18-ft. length of the 2x12 ridge on the extended rafter section, I was hoping against hope to put the 2x4 ridge block at the new gable rafters dead in my sights. It was a clear miss. The block was 1½ in. off toward the back of the house. The inch I was off at the mudsill had followed me up to the ridge, and it picked up another ½ in. to boot. Not good, but not disastrous either. I had to stick with the gable rafters I'd already nailed in place so I could keep a flat plane on the front roof; I couldn't cheat here. But I could on the dormer side of the ridge. I decided to sister a new 2x10 ridge onto the face of the 2x12 ridge of the extended rafter section and run it all the way to the far gable end. This put the new ridge in perfect alignment with the new gable rafters. Thus, the front rafters all worked out to be the same length, but each of the dormer rafters had to be cut a tad shorter than the previous one as they progressed toward the gable end.

Framing up a shed dormer and introducing

a vertical wall where the rafters ordinarily go means losing the strength and rigidity of a tri-angulated structure. If something isn't done to compensate for this loss, the ridge will sag and the dormer wall will lean out of plumb—if you're lucky. In the worst instance, the ridge (which, with dormer rafters attached, be-comes a structural member) can fail and cause the roof to collapse. I solved the prob-lem in this case by framing up a truss-like ar-rangement that consists of a long 2x10 rafter, a short 2x8 rafter, a 2x8 bottom chord, which functions as a collar tie and serves as a ceiling joist, and a 1x8 hanger, which supports the joist about ⅓ of the way along its 21-ft. span. This 1x8 looks a little like a king post, but its purpose is not to take any compression load-ing. This restores the triangulation you lose in omitting the rafters and framing up a dormer wall. (In most shed-dormer situations, where joists aren't spanning long distances, the hanger isn't needed.) The joists sit on the double top plate of the dormer window wall, where they extend to the plumb cut on the rafter tails and are well nailed to each rafter.

So each joist would be at the right height when I nailed it to its hanger, I cut a 7½-ft. long 2x4 for a gauge stick. The joist would sit on top of the gauge stick while I nailed the hanger to the joist. To stiffen the joists and to help keep them from twisting and sagging, I ran a *strong back* (detail drawing, above right) across them in the middle of their span be-tween the hangers and the dormer wall.

Because at this point I wasn't ready to re-

move the old roof under the dormer, I couldn't yet install the 2x8 ceiling joists in the 18-ft. long section of roof that was over the old roof. But that part of the structure did need stabilizing, so I nailed a temporary 1x8 tie to each rafter at the dormer wall plate and angled it upward to catch the extended rafter opposite, just above the plate at the ridge. These ties would be replaced with permanent 2x8 ceiling joists to match those in the rest of the dormer, once I built an interior 2x6 stud wall to support the old ridge. Things were now secure enough for me to remove the ridge supports.

The roof framing on the addition was com-plete at this point, except for the short eave rafters that would extend from the dormer window wall down to the double top plate of the wall below. Instead of measuring these rafters on my deck layout, I picked the correct length for them by measuring them in place. These short 2x10 rafters, at 16 in. o. c., have the same bird's-mouth cut at the bottom as the full-length rafters. They were nailed at the top to the 2x8 ledger on the face of the dor-mer window wall.

After all the roof trim was on, I shingled the whole roof with GAF Timberline fiberglass shingles. Their texture doesn't show the im-perfections that are bound to occur when a new roof meets an old one. A flat shingle lets every bump and depression from the decking below telegraph through.

The old roof section under the new dormer roof could now be opened without fear of wa-

Illustrations: Christopher Clapp

Framing the dormer. The long window wall and the dormer cheeks are set back 4 ft. from the walls below, allowing the original roof plane to show on all sides (photos at right). At this stage, the dormer window wall and cheeks have been framed and sheathed. A ledger board along the bottom of the window wall carries the short rafters. The plumb cuts on the rafter tails have been made flush with the plate, and dummy tails applied, first to a long 1x board, and then to the ends of the true rafters. This makes the rafter tails line up, the fascia and soffit easier to apply, and it eliminates tedious blocking.

ter damage. I stripped the shingles and plywood decking from the old roof section, but left the rafters in place for the time being. Next, I cut and nailed the permanent ceiling joists in this section in place, and took care to level them with the rest of the joists in the dormer. Instead of nailing their inboard ends to an interior wall plate, as I had done with the other joists, I face-nailed them to the existing rafters after mitering their ends at the roof pitch to fit snugly under the deck. Then I reinforced the joists with 1x8 hangers and a strong back, just as I had done in the rest of the dormer. Once these were installed, I removed the 1x8 ties that had held the wall and roof together.

Before removing the old rafters that I had just exposed, I still had to take care of the one disturbing potential weakness in this framing scheme. The juncture at which the 7-ft. extended rafters sit on the plate at the old ridge needed some support, or the roof might sag at that point. I wasn't afraid that the roof would actually collapse, but I did want an extra measure of assurance.

I have taken apart lots of old houses, and my experience and intuition told me that it would be worth the extra time to give this weak link a little extra support against that unpredictably heavy snow load. To do this I built an interior stud wall of 2x6s that sits directly under the ceiling joists where they tie into the old rafters, as shown in the drawing on the facing page. Because this wall had two wide closet doors and a skylight in its 18-ft. length, I notched a double 2x10 header into the 2x6 studs to carry the load of the ceiling joists and rafters above. To get good bearing below, I had to beef up the existing beam that had been installed by a previous remodeler because I doubted its ability to support the increased load.

This framing system had, of course, all been carefully planned out ahead of time, using past experience and common sense. I knew it would work. So when I tore out the old gable-end wall and cut away the old rafters leaving only stubs behind, I didn't bat an eye. Just kept my fingers crossed.

I'm very pleased with the way this job turned out, and as a bonus, Jasper's cache was discovered in a roof gutter across the street. I now have a good supply of nail sets and drill bits, and an extra set of car keys. □

Bob Syvanen is a consulting editor with Fine Homebuilding *magazine. Photos by the author.*

The finished house, seen from the front, appears smaller than it actually is because the extended roofline obscures the new 40-ft. shed dormer at the back of the house.

Shed-Dormer Retrofit

Turning your attic into living space may be the remedy for your growing pains

by Scott McBride

Growing up amid the post-war baby plantations of central Long Island, I got to see a lot of expand-as-you-go housing. One of my earliest memories is the sight of slightly dangerous looking men, with hairy arms and sweaty faces, tearing the roof off of our home. My parents had decided to add onto our modest Cape, and that meant building a shed dormer. The following spring, a neighbor came over to take measurements; his house and ours, you see, were identical, and he wanted to do the same thing to his place. Before long, all the houses in our subdivision had sprouted the same 14-ft. long dormer.

Rivaled only by the finished basement, the enlarged and finished attic endures as the most practical way for the average suburban family to ease its growing pains. The shed dormer makes it possible to enlarge almost any attic space simply by flipping up the plane of the gable roof. Compared to the cost and complexity of a gable dormer, the shed dormer is a good choice where size and budget take precedence over looks.

Design—Shed dormers may be so narrow as to accommodate only a window or two, or may run the entire length of the house. In the latter case, it is common to leave a strip of the main roof alongside the rake at each gable end (photo above).

The trickiest part of designing a dormer is getting the profile right. To find the correct position of the inboard header and the dormer face wall, begin by making a scale drawing of the existing roof. Then draw in the dormer that you have in mind. What you're trying to determine here are the height of the dormer's face wall, the pitch of

its roof, and where these two planes will intersect the plane of the main roof.

When determining the height of the face wall, consider exterior appearance, interior headroom, and window heights. The roof pitch you choose will affect the kind of roofing. Shingles require at least a 4-in-12 pitch. A flatter pitch should be roofed with 90-lb. roll roofing. This usually isn't a visual problem because you can't see the flatter roof from the ground.

Loading and bearing—Once you're sure that the existing ceiling joists will support live floor loads, you have to consider the other structural aspects of adding a shed dormer. Removing all or part of the rafters on one side of a gable roof upsets its structural equilibrium. You're taking a stable, triangulated structure and turning it into a not-so-stable trapezoid. The downward and outward forces exerted by the remaining rafters are no longer neatly countered by opposing members. The dormer's framing system has to compensate for this lost triangulation. To understand how this happens, let's take a look at a dormer's structural anatomy. As shown in the drawing on the facing page, the *inboard header* transfers loading from the *cripple rafters* out to the *trimmer rafters* on either side of the dormer. The full-length trimmer rafters send this lateral thrust down to the joists. With the main roof load reapportioned around the dormer, the new roof is structurally able to stand on its own.

On low-pitched dormers, the roof sheathing acts as a sort of horizontal beam that reinforces the inboard header and helps transfer the lateral thrust of the main roof out to the trimmer raf-

ters. As you increase the pitch of the dormer, you decrease the ability of the dormer roof to act as a horizontal beam. And the lateral force of the dormer rafters themselves will sometimes threaten to bow out the dormer face wall. The solution is to tie the main-roof rafters and dormer rafters together with ceiling joists. These act as collar ties, creating a modified version of the original gable triangle.

We also have to consider vertical loads. The dormer roof on the outboard side is supported by the dormer face wall, which is built either directly atop the exterior wall or slightly to the inside, where it bears on the attic floor joists. This second option, shown in the drawing, facing page, leaves a small section of the original roof plane (called the apron) intact, and lets you retain the existing cornice and gutter. It also sets the dormer back a bit from the eave line and visually reduces the weight of the addition. Depending on the size of your floor joists, if the setback is more than one or two feet, the load on the attic floor joists can become too great. To lighten this load, you should install a header at the top of the apron to carry the roof load out to the trimmer rafters. In any case, this face wall will support a little more than half the weight of the dormer, depending on the roof pitch.

The other half of the dormer roof load usually rests on a large inboard header, which transmits the load through the trimmer rafters down to the exterior walls. To increase roof pitch and gain more headroom, the inboard header is frequently moved all the way up to the ridge of the main roof. If this ridge beam is made strong enough to carry roughly half the weight of the dormer

From *Fine Homebuilding* magazine (October 1986) 35:46-50

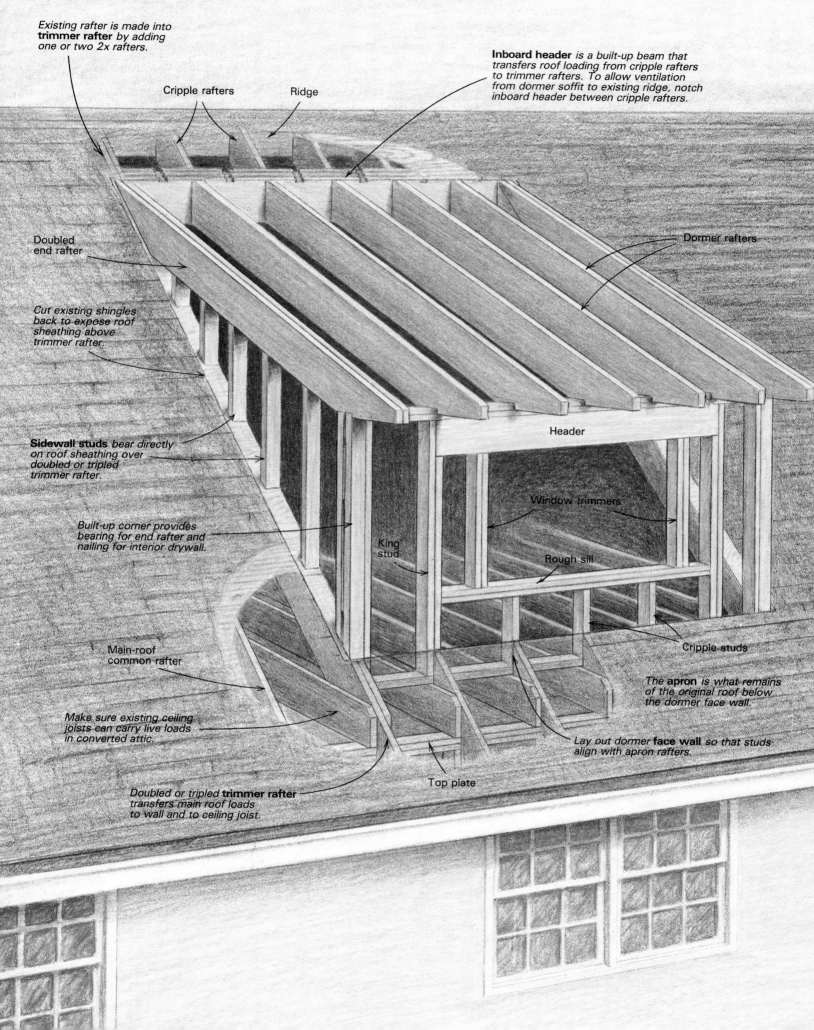

Retrofit framing details

Existing rafter is made into **trimmer rafter** by adding one or two 2x rafters.

Cripple rafters

Ridge

Inboard header is a built-up beam that transfers roof loading from cripple rafters to trimmer rafters. To allow ventilation from dormer soffit to existing ridge, notch inboard header between cripple rafters.

Doubled end rafter

Dormer rafters

Cut existing shingles back to expose roof sheathing above trimmer rafter.

Header

Sidewall studs bear directly on roof sheathing over doubled or tripled trimmer rafter.

Window trimmers

Built-up corner provides bearing for end rafter and nailing for interior drywall.

King stud

Rough sill

Main-roof common rafter

Cripple studs

The **apron** is what remains of the original roof below the dormer face wall.

Make sure existing ceiling joists can carry live loads in converted attic.

Lay out dormer **face wall** so that studs align with apron rafters.

Doubled or tripled **trimmer rafter** transfers main roof loads to wall and to ceiling joist.

Top plate

roof and half the weight of the main roof, then it won't sag, and the rafters connected to it cannot spread apart at the plates. This allows the attic to have a cathedral ceiling.

If the ceiling is to be flat, the ceiling joists will prevent the roof from spreading, as mentioned earlier. In this case the ridge is non-structural and can be made of lighter stuff.

If you don't use ceiling joists and go for a cathedral ceiling, the length of your dormer will depend upon the practical length of the inboard header or structural ridge beam. About 12 ft. to 16 ft. is typical. At this length, a triple 2x10 or 2x12 should make an adequate header, capable of carrying half the dormer roof load, plus the weight of any cripple rafters above it. If the 2xs in the built-up header are slightly offset from one another and the main roof pitch is steep

enough, the header will not protrude below the ceiling. Sizes of all members should be checked by an engineer, architect or building inspector.

If you're going to build a long dormer, you can support the header between the trimmer rafters with an intermediate rafter. Hidden inside a partition wall that runs perpendicular to the face wall, this rafter picks up the load of the headers, which can then be reduced in size.

Preparation—Before you cut a big hole in your roof, you have to determine the location of the dormer from inside the attic. Lay some kind of temporary floor over the open joists to keep boots from going through ceilings and to keep trash out of the attic insulation.

You may want to use one of the existing rafters as a starting point and lay out the dormer from

there. In this case the existing rafter becomes a trimmer rafter and will have to be doubled or possibly tripled to carry the load. This can be done before the roof is opened up.

To lay out these extra trimmer rafters, measure the underside of an existing rafter from the heel of the plumb cut at the ridge down to the heel of the level cut at the plate. Transfer the respective angles with your T-bevel. These extra rafters don't support the cornice, so you don't have to cut a bird's mouth; just let the level cut run through. If a ceiling joist prevents the new rafters from reaching the plate, raise the level cut on the bottom of the rafters so they bear snugly on the top edge of the joist.

Now slide the additional rafters into the appropriate bays to make the trimmers. Any roofing nails protruding below the sheathing should

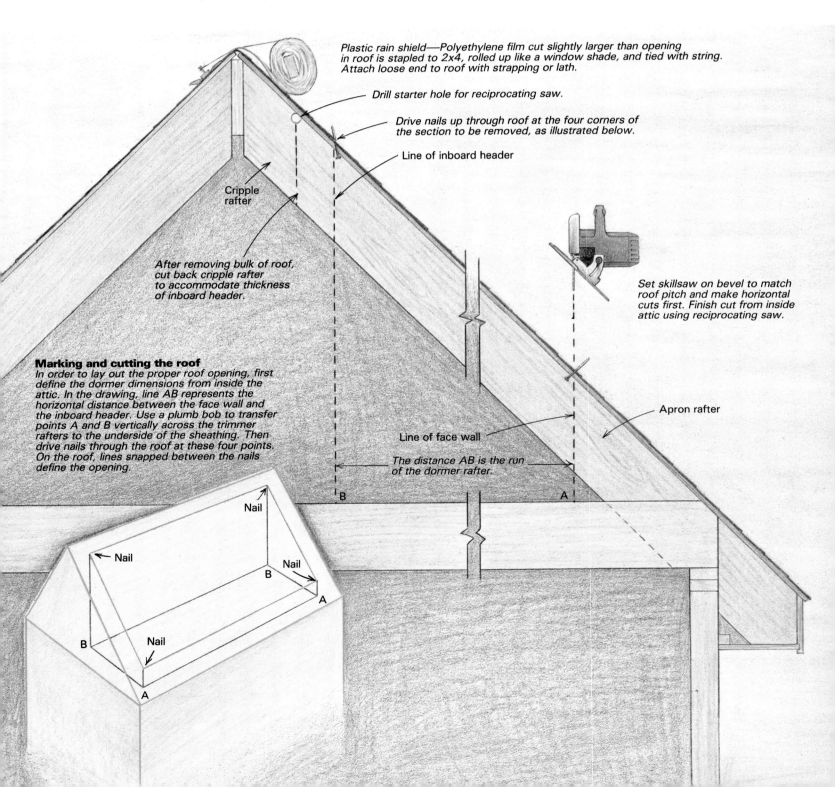

Plastic rain shield—Polyethylene film cut slightly larger than opening in roof is stapled to 2x4, rolled up like a window shade, and tied with string. Attach loose end to roof with strapping or lath.

Drill starter hole for reciprocating saw.

Drive nails up through roof at the four corners of the section to be removed, as illustrated below.

Line of inboard header

Cripple rafter

After removing bulk of roof, cut back cripple rafter to accommodate thickness of inboard header.

Set skillsaw on bevel to match roof pitch and make horizontal cuts first. Finish cut from inside attic using reciprocating saw.

Marking and cutting the roof
In order to lay out the proper roof opening, first define the dormer dimensions from inside the attic. In the drawing, line AB represents the horizontal distance between the face wall and the inboard header. Use a plumb bob to transfer points A and B vertically across the trimmer rafters to the underside of the sheathing. Then drive nails through the roof at these four points. On the roof, lines snapped between the nails define the opening.

Line of face wall

The distance AB is the run of the dormer rafter.

Apron rafter

B   A

Nail

Nail   Nail
              B

B   A

Nail

A

be nipped off. Some persuasion may be necessary to bring the new rafter up tight against the old one. Spike the rafters together generously and toenail the new ones to the ridge and plate.

Now that you've defined the length of the dormer, you need to mark off the width. Begin by measuring from the outside of the exterior wall in to where you want the face wall (drawing, facing page). From here plumb a line up across the trimmer rafter and mark where this line intersects the roof. Now measure horizontally toward the ridge, from the proposed face wall to where you want the inboard header. Plumb another line up to the trimmer from here, and mark where it intersects the roof. This distance is the width of your dormer; it's also the run of the dormer rafter. Where these points (two on each set of trimmer rafters) touch the underside of the sheathing, drive four large nails up through the roof to mark the corners of the rectangular section you'll cut out from above. But before heading up to make the cuts, check for electrical wires, vent stacks and anything else you don't want your skillsaw to run into.

Rigging—Since houses with steep roof pitches make the best candidates for dormers, you'll need good rigging. Set up staging along the eaves, extending a few feet past both sides of the dormer location. If a hoist or pulley can be rigged in conjunction with the scaffold, so much the better. To gain access along the sides of the dormer, a ladder can be hooked over the ridge, or roof brackets can be set up.

"What happens if it rains?" is the question most often asked by clients. If proper precautions are not taken while the house undergoes dormer surgery, a heavy rainstorm could cause thousands of dollars in damage.

Once you get up to the ridge, install an emergency rain shield—a piece of heavy polyethylene film wrapped around a 2x4 somewhat longer than the length of the dormer. On the ground, spread out a piece of the poly several feet longer than the dormer and wide enough to reach from the ridge of the existing roof to the eaves. Staple one of the horizontal edges to the 2x4, and roll the sheet up like a window shade. Tie the roll with string, then carry it up to the roof and fasten its free edge to the ridge with wood lath or strapping. If it rains, cut the string and let the sheet unroll. The weight of the 2x4 hanging over the eaves will keep the poly tight, so it won't flap in the wind and so puddles won't develop.

Demolition—For cutting through the roof, you need a powerful skillsaw equipped with a nail-cutting blade. The heavy carbide tips on these blades are ground almost square, giving them the toughness needed to plow through asphalt, plywood and miscellaneous nails all at once. Some manufacturers coat this type of blade with Teflon to reduce friction. Eye protection is a must during this operation.

After snapping lines between the four nails you drove up through the roof, make the horizontal cuts first (there are only two of them and they're a little tougher to do). Set the skillsaw as deep as it will go, and set the saw's shoe to the plumb-cut angle of the main roof. Since you cannot safely plunge-cut with a skillsaw when it's set on an angle, start the cut with your drill and reciprocating saw. Then use a slow, steady feed on the skillsaw. Keep moving, because the weight of the saw will tend to push the downhill side of the sawblade against the work, generating extra friction. Be particularly alert to the possibility of kickback; your blade will be crashing into 8d sheathing nails now and then. And remember, you're up on a roof.

To make the vertical cuts, set the skillsaw back to 90° and start the cut on the uphill end. These cuts are easier because the weight of the saw helps pull it through the cut. All you have to do is slide down the roof behind it. If you have had experience plunge-cutting, then begin this way. Otherwise, start the cut with the reciprocating saw, and finish with the skillsaw.

After the four outline cuts are done, make longitudinal cuts down the middle of each bay in the area to be removed. This divides the roof into manageable chunks. Before freeing these chunks by completing the cuts through the rafters, determine whether the remaining roof frame (the apron rafters below, and the cripple rafters above) need temporary support. If these pieces are short and well-nailed to the plate and ridge, they will stay up by themselves. If not, shore them up temporarily with 2x4 braces.

Now use your reciprocating saw inside the attic to complete the cuts through the rafters. Have a couple of burly helpers hold up each section while you're working on it. A 10-ft. 2x8 rafter, 14 sq. ft. of sheathing, and several layers of roofing make these chunks very heavy. The safest way to lower them to the ground is with a strong rope that's wrapped around a sturdy mast.

After removing the bulk of the roof, the bottom ends of the cripple rafters must be cut back to accommodate the thickness of the inboard header, without cutting through the sheathing. Drill a ¾-in. starter hole at the top of the mark with a right-angle drill. Then cut straight down with the reciprocating saw.

To finish up the demolition, you'll need to cut back the roofing material to make way for the dormer sidewalls. Set the depth of the skillsaw so that it will cut through all the roof shingles, but will just graze the sheathing. Snap longitudinal lines on the roofing, located back from the inside faces of the trimmers a distance equal to the width of the dormer sidewall framing plus sheathing thickness, plus ½ in. for clearance. Slice through the roof shingles along these lines, and peel back the roofing to expose strips of decking above the trimmer rafters. The dormer sidewalls will be built up from these, with the inside edges of the studs flush with the inside faces of the trimmer rafters.

Wall framing—Your next step will be to cut and lay out the plates for the face wall. In most situations, the bottom plate for the face wall will bear directly on the attic floor. You should lay out the face-wall framing so that the apron rafters will bear directly on the wall studs (aligning the framing in this way is called stacking).

In order to bring concentrated roof loads down safely onto the floor framing, window king studs also should be in line with the apron rafters or else be located over a joist. You can then frame inward to get the necessary rough-opening width. Or you can forget all this and just double the bottom plate to distribute the load safely.

Taking the scaled measurements from your drawings, transfer the header and sill lengths onto the plates. The various stud lengths will also come from your drawings. Be sure to locate the rough sill for the windows at least several inches above the apron to keep rain and melted snow from creeping in underneath.

Cut all the face-wall components and assemble them on the attic floor. Then raise and plumb the wall, bracing it temporarily if necessary.

Next you have to fill out the corners of the face wall with a combination of beveled sidewall studs and blocking, as shown in the drawing on p. 105. The tops of these studs will be flush with the top of the face wall, and will give bearing to the dormer end rafters. Begin by cutting oversized pieces with the pitch of the main roof cut on one end. Stand these in place and mark their tops flush with the top of the face wall. Cut and nail. This completes the face wall.

Roof framing—The shed-dormer rafter is laid out just like any common rafter. The only differences are the generally lower pitch, and the fact that its plumb cut bears against a header instead of a ridge board. There are several ways to determine rafter length. I lay out the bird's mouth first, then step off the rafter length with a large pair of dividers, using the method I described in my article on roof framing (pp. 85-91). After marking the plumb cut at the ridge, lay out the rafter tail according to the soffit and fascia details from your elevation drawings. Then carefully cut out the rafter pattern. (For more on roof framing, and a glossary, see pp. 82-83.)

Turning to the roof, first nail the inboard header to the trimmers. Joist hangers won't work in this situation because you'll want to offset the 2xs like stair steps, starting each one slightly above or below the next in order to fit the slope of the roof. Instead, just toenail each piece in place with plenty of 16d commons, and then spike them to each other.

Now try the rafter pattern at several different locations along the top plate of the face wall. If all is well, use the pattern to cut the rest of the rafters and nail them in place. The spacing of the rafters should align with the face-wall studs in the same way the studs align with the floor joists. The end rafters will have to be retrimmed on a sharp angle at the top, because they bear directly on the roof instead of on the header. Place one of the pattern-cut rafters in position, up against the trimmer rafter, and mark the roof-line. After cutting, double the end rafters to provide nailing for drywall, or spike a 2x on the flat with its bottom face flush with the bottom edge of the end rafter.

The dormer sidewall studs are framed directly from the main roof up to the dormer's end rafters, without any plates. As with the corners of the face wall, begin by cutting oversized pieces with the pitch of the main roof cut on one end. Then stand the pieces in place and mark where they meet the dormer end rafter,

Piggyback shed dormers. Shed dormers are often part of the original design on houses that have a gambrel roof. Here a second dormer was added on top of the first, probably to let more light into the room. The piggyback dormer also accommodates an air conditioner.

cut along the mark and nail them up. These studs diminish in a regular progression, like gable studs, as they approach the ridge. If you don't want to mark each one in place, just mark the first two and measure the difference between them. This measurement is their common difference, and you can use it to calculate the diminishing lengths of the remaining studs.

Closing up—A few points on exterior finish are worth mentioning. Before decking the dormer roof, use a shingle ripper to remove nails in the first course of roof shingles above the dormer. This allows the dormer roofing material to be slipped underneath the existing shingles. If you install the sheathing first, the lower pitch of the dormer roof will interfere with the handle of the shingle ripper.

Flashing a shed dormer is relatively simple. As shown in the drawing below left, the apron is flashed first, then the dormer sidewalls are step-flashed. Use a 6-in. wide length of flashing along the apron, creased in half so that 3 in. of flashing runs up the dormer face wall and 3 in. extends over the apron shingles. Nail the face-wall side only. At the corners of the face wall, let the apron flashing run a few inches past the dormer sidewall. Slit the flashing vertically along the corner of the dormer, and push the overhanging vertical fin down flat on the main roof.

Overlap the apron flashing with the first piece of step flashing, where the dormer sidewall meets the main roof. Extend the step flashing down at least an inch past the face wall, and fold the vertical fin down and back on an angle. This will carry rainwater safely past the corner. You'll have to relieve the back of the corner board to fit over this first piece of step flashing.

Continue the step flashing all the way up the sidewall, slipping one piece of bent step flashing under the end of each roof shingle course, and pressing the other side up against the wall sheathing. Don't nail the step flashing into the roof; nail it to the sidewall only.

The rake board is usually furred out with a piece of 5/4 spruce so the siding can be slipped underneath. Where the dormer rake board dies into the main roof, the uppermost piece of step flashing is trimmed on an angle so that it can fit up behind the rake board and tight against the furring. Give the dormer roofing a little extra overhang here to help divert water from this sensitive spot.

Vents in the dormer soffit are a good idea. They prevent condensation in the dormer roof insulation as well as ice damming at the eaves. Since the inboard header blocks the flow of warm air at the tops of the roof bays, cut some notches across the top of the header in each bay or recess the top edge of the header slightly below the tops of the dormer rafters. If you're insulating between the rafters (instead of the ceiling joists), you'll need some spacers to create an airflow channel between the roof sheathing and the insulation. This allows some air flow from the dormer soffit vent to the ridge vent or gable-end louvers. □

Scott McBride is a carpenter and contractor in Irvington, N. Y.

Roofing and flashing details

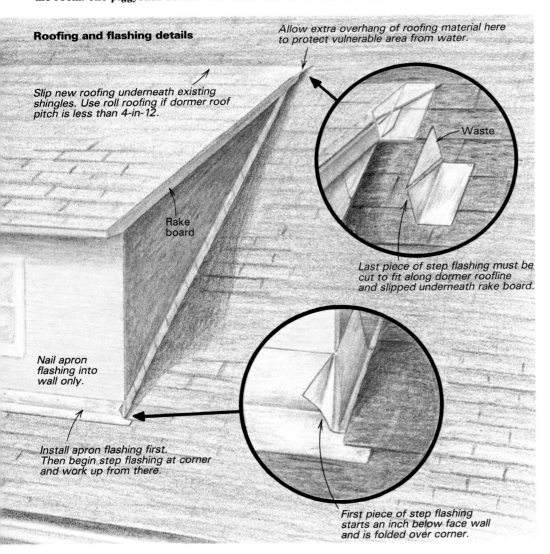

Allow extra overhang of roofing material here to protect vulnerable area from water.

Slip new roofing underneath existing shingles. Use roll roofing if dormer roof pitch is less than 4-in-12.

Rake board

Waste

Last piece of step flashing must be cut to fit along dormer roofline and slipped underneath rake board.

Nail apron flashing into wall only.

Install apron flashing first. Then begin step flashing at corner and work up from there.

First piece of step flashing starts an inch below face wall and is folded over corner.

Photos: Scott McBride

Peaking Over a Flat Roof

Analyzing new loads on an old structure

by Max Jacobson and Murray Silverstein

Sometimes building up is the most logical way to remodel, as we discovered in designing an addition for a small house in northern California. The existing house was essentially a developer-built plywood box with an entrance deck and living spaces on the upper level and bedrooms below. The site, a relatively remote slope in the Oakland, California hills, is a dramatic spot surrounded by large pine trees, with a view of the Bay and the Golden Gate to the west and a canyon to the northeast.

The drama of this project comes from the way a relatively simple act of remodeling can so completely transform a house—indeed, the very image of home. In this project, the owners had been attracted to the site, a secluded spot on a hill high above the city, far more than to the house. These were people who spent very busy, hectic days in the city, and who were looking for a place where they could spend quiet, relaxed nights outside and above the city.

The house did not fit the site, and it looked like a rather unsheltering shoe box; the flat roof contributed nothing to the feeling of shelter inside or out. Indeed, since the house was approached from above, along a winding driveway, the flat gravel roof was one of the most visible elements of the building. The best solution to this problem seemed to be the addition of a gable roof over the house. The steep lines of the roof would be at home among the tall trees; the roof would be visible from around the canyon. Within its volume the roof could contain additional rooms and inside it could be visible all the way up to its ridge through the clerestory from the main level at entry grade below.

Working from a model, we decided that a 48° roof pitch gave us the form we wanted. It was sufficiently visible and evocative, and the steep pitch would allow us to create four new spaces under it—two rooms, a desk alcove and a tiny greenhouse—and still maintain a 10-ft. by 10-ft. opening to the main level.

Structural design—Our structural engineer was not happy with the decision to add such a large, new roof to what he considered a rather shoddily built old house. The new roof would add enormous wind and earthquake loads to the old structure that it was never designed to resist. As we worked through the structural design process with the engineer, we learned a lesson in analyzing new loads on old structures. When one designs a vertical addition to an existing building, the resulting structure must comply with California's local building codes, just as if the entire structure were being built from scratch. You cannot simply assume that if the old structure is up to code, and you build on top of it using conventionally adequate construction, that the whole ensemble will be structurally sound. In this case, three new factors had to be considered: Were the original foundations big enough to support the weight of a new building? Would the entire structure, with the new addition, be stiff enough to transfer increased wind or earthquake loads to the foundations? If the answers to these questions were "yes," would the building be able to resist overturning in the face of increased wind loads?

The vertical loads that must be carried by the foundation consist of both the weight of the empty building itself (the dead load) and the weight of the occupants, their belongings, and the snow or rain that sits on the building (the live load). Since the weights of building materials are known, we simply add up the various quantities.

Continued on p. 111.

The new balcony overlooks a 10-ft. by 10-ft. clerestory to the main floor. Wire-glass skylights face north to allow soft illumination. The finished addition dramatically alters the appearance of the once flat-roofed house. Redwood 2x10s envelope protruding beams to hide the rafter connection.

From *Fine Homebuilding* magazine (February 1981) 1:40-42

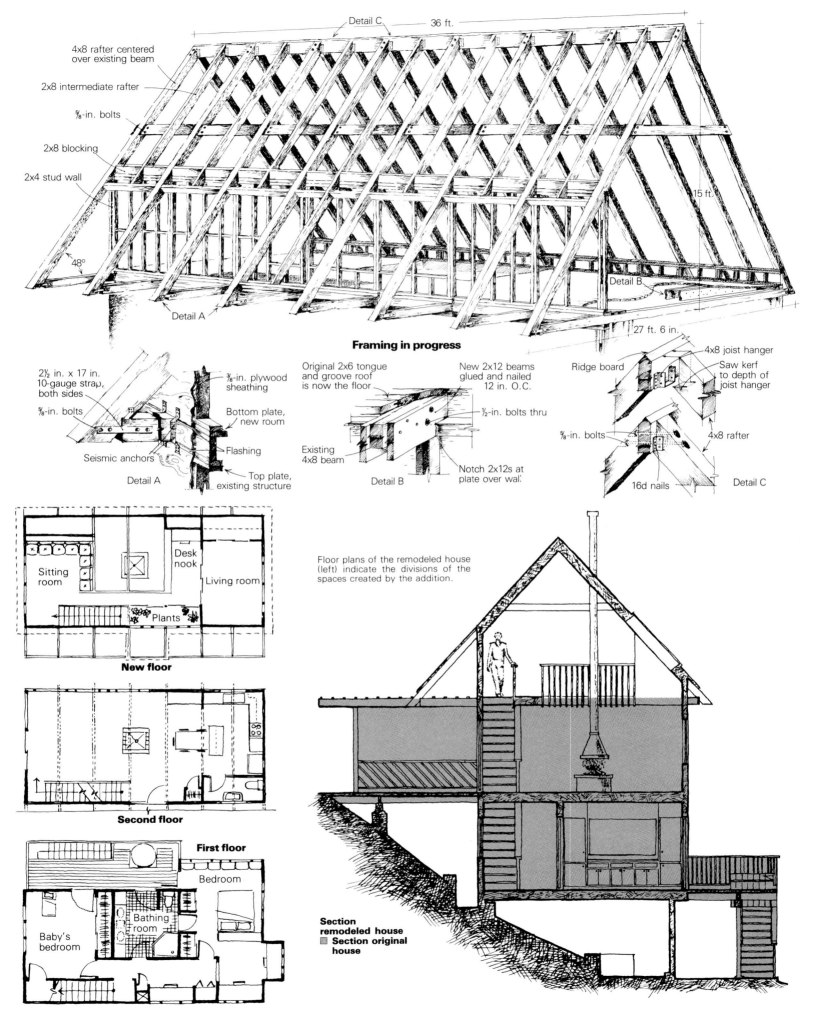

Detail C — 36 ft.

4x8 rafter centered over existing beam

2x8 intermediate rafter

⅝-in. bolts

2x8 blocking

2x4 stud wall

48°

Detail A

15 ft.

Detail B

27 ft. 6 in.

Framing in progress

2½ in. x 17 in. 10-gauge strap, both sides

⅝-in. bolts

Seismic anchors

Detail A

⅞-in. plywood sheathing

Bottom plate, new room

Flashing

Top plate, existing structure

Original 2x6 tongue and groove roof is now the floor

New 2x12 beams glued and nailed 12 in. O.C.

½-in. bolts thru

Existing 4x8 beam

Notch 2x12s at plate over wall.

Detail B

Ridge board

4x8 joist hanger

Saw kerf to depth of joist hanger

⅝-in. bolts

4x8 rafter

16d nails

Detail C

Sitting room

Desk nook

Living room

Plants

New floor

Second floor

First floor

Bedroom

Baby's bedroom

Bathing room

Floor plans of the remodeled house (left) indicate the divisions of the spaces created by the addition.

Section remodeled house
▨ **Section original house**

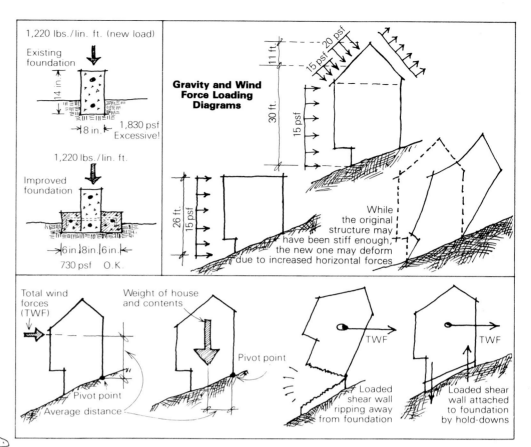

Original ⅜-in. plywood with added nails

New ½-in. plywood on interior wall

Floor plate

Double 2x10 joist

1-in. threaded rod

4x4 stud

1 in. bolts

¼-in. steel hold-downs

2x4 mudsill

Existing foundation

New foundation

½-in. steel dowel

½-in. rebar

1,220 lbs./lin. ft. (new load)

Existing foundation

14 in.

8 in.   1,830 psf Excessive!

1,220 lbs./lin. ft.

Improved foundation

6 in. 8 in. 6 in.

730 psf O.K.

Gravity and Wind Force Loading Diagrams

20 psf

15 psf

11 ft.

30 ft.

15 psf

26 ft.

15 psf

While the original structure may have been stiff enough, the new one may deform due to increased horizontal forces

Total wind forces (TWF)

Weight of house and contents

Pivot point

Average distance

Pivot point

TWF

Loaded shear wall ripping away from foundation

TWF

Loaded shear wall attached to foundation by hold-downs

Wall studs are attached directly to the foundation by steel hold-downs. In this way, walls, existing foundation and new foundation work together to resist overturning forces.

Similarly, the code specifies reasonable estimates of the load generated by people and furnishings on the floors, as well as the loads generated by snow on the roof. (Since steeper roofs will carry less snow, the code adjusts the estimated load for the pitch of the roof.) The dead and live loads must be distributed through the structure down to the foundation which, in turn, pushes down upon the underlying earth. Normally this bearing pressure should not exceed 1,000 lbs. per sq. ft. In our project, with our new floor and roof, the pressure would have been almost twice this figure, making it necessary to double the old foundation's bearing area.

An existing foundation can be enlarged by placing new concrete adjacent to the original. In this case, connecting the new with the old was accomplished by inserting ½-in. diameter steel dowels through ⅝-in. holes in the original concrete. These dowels must protrude into the new concrete far enough to overlap the steel reinforcing rods in the new foundation.

Lateral forces generated by earthquakes and wind were considered next. In the case of wind, a bigger building will experience a larger force simply due to its increased area (force = wind pressure × area). In addition, the wind pressure on the building increases as the height above the ground increases, from around 15 lbs. per sq. ft. for the first 30 ft., to 20 lbs. per sq. ft. for the next 30 ft., and so on. Also, the addition of a pitched roof to an existing flat-roofed building brings new forces into play, acting on both the windward and leeward sides.

While the original structure may have been stiff enough, the new one may deform due to the increased horizontal forces. Ultimately, these forces must be transfered into the walls parallel to the wind direction under consideration, and these walls must be resistant enough to transfer their forces down to the ground. Often this will entail the addition of plywood stiffeners, or increased nailing of the existing plywood walls, along with better attachment of these newly stiffened walls to their foundations with hold-downs. Hold-downs are devices which enable the wall studs to be bolted directly to the foundation, as opposed to the conventional connection of toe-nailing to the plate.

Finally, consider whether the entire vertically loaded structure—roof, floors, walls and foundation—can resist overturning in the face of lateral loads (such as heavy winds). The designer will choose some arbitrary pivot point on the building (a corner, for instance) and check that the tendency of the wind to push the building over in one direction is not greater than the tendency of the vertically loaded building to push back in the opposite direction. The total wind forces multiplied by their average distance from the pivot point must be less than the total weight of the building multiplied by the average distance from the same pivot. Fortunately, the added foundation mass (required earlier to reduce bearing pressures) helped increase the building's resistance to overturning by increasing the building's weight. □

Max Jacobson and Murray Silverstein are architects, teachers and writers in Berkeley, Calif.

Framing Doghouse Dormers

Two ways to frame a basic gable dormer

by Scott McBride

I think of the dormer as one of the more playful aspects of a house. It wants to poke its head up and make a little mischief with the roof. The design of a dormer should echo the house's main roof in spirit, but not necessarily in detail. The Victorians enjoyed punctuating their roofs with all sorts of crazy outcroppings, and their adventurous spirit seems sadly missing from much of today's architecture.

The word "dormer" comes from the Latin verb *dormire,* to sleep. But while this suggests the dormer's function inside the house—to admit light and air into an attic bedchamber—it gives no indication of the aesthetic possibilities of this versatile architectural feature.

Dormers have been adapted for use on just about every style of house with a pitched roof. Although a few miss the mark, most succeed in lending some measure of character and charm to a home's appearance.

Gable dormers, because of their small scale, represent a microcosm of roof-framing theory and practice. And as such, they provide a good opportunity for novice carpenters and builders to study this complicated subject.

Walls—The basic gable dormer has a rectangular face wall and two triangular sidewalls, also known as "cheeks." The face wall is usually built up from the attic floor, or from an outboard header in the main roof, and the sidewalls are framed up from trimmer rafters in the main roof. (For more on dormer wall framing, see pp. 104-108.)

As a general rule, the face wall of a gable dormer should be mostly window, with little or no siding on either side. A dormer should wink at you—not sit on the roof like a refrigerator with a mail slot. Bring the window rough opening right out to the corner posts. Since the gable end is non-bearing, you can usually omit the window jacks, and let the top plate define the height of the rough opening. Frequently, the corner board and exterior window casing are one piece, with solid trim covering the gable as well. This eliminates any siding on the face wall.

The triangle above a gable dormer window is traditionally a place where carpenters love to flaunt their woodworking skills. Appliqué, fancy-cut shingles and decorative truss work (stick style) are just a few of the treatments found here. Houses in the more formal Georgian and Adam styles often use the gable to field an elegant half-round fan window.

The exterior finish of the sidewalls can be either siding material or roofing material, according to taste. Slate looks exceptionally good here, as do handsplit shakes.

Gable-roof styling—Getting the right pitch on the dormer roof is essential. You can do your planning on paper, but it's a good idea to mock up the roof lines after the walls are framed, either with 1x4s or by cutting out the dormer profile in a sheet of plywood. Since the dormer will usually be viewed from below (not straight on) elevation drawings are of limited value in judging its appearance. Whenever possible, use the main roof pitch for the dormer roof as well. This saves a lot of headaches in the framing.

Roof framing—The trickiest part of dormer construction is framing the intersection (valley) of the dormer roof and main roof. The type of ceiling inside the dormer determines which framing method to use. There are two choices.

The simpler approach, which I call the valley-board method, is to build the dormer roof on top of the main roof (drawing, facing page, left). Doing it this way means having a flat ceiling below the dormer because the main-roof cripple rafters cut off the dormer roof space from the rest of the attic.

I call the second approach the valley-rafter framing method. It has the inboard header set at the elevation of the dormer ridge, rather than at the level of the wall plates (drawing, facing page, right). This allows for a cathedral ceiling inside the dormer. It is more trouble than the valley-board method, and requires a more thorough knowledge of roof-framing geometry.

Valley-board method—I'll begin with the simpler of the two approaches. After framing the dormer walls, set the inboard header (usually a pair of 2x6s or 2x8s) between the trimmer rafters with its bottom edge flush with the top of the dormer sidewall plates. This way the dormer ceiling joists will line up with the header.

Next, fill in the main-roof frame between the trimmer rafters by installing the cripple rafters, which extend from the inboard header up to the main roof ridge. These cripples have a plumb cut on both their upper and lower ends. If the cripples are long, it's a good idea to notch the lower ends so they hook over the header. This brings the weight of the cripple to bear on the top of the header. Otherwise, the strength of the joint depends solely on nails in shear. Metal framing connectors (sloped-seat joist hangers) can also be used to reinforce this connection.

If you sheathe the main roof before framing the dormer roof, you'll avoid the difficulty of cutting plywood around the dormer later. But you may want to leave the framing open to the dormer attic space in order to ventilate the roof bays and keep the insulation dry. In this case, go ahead and sheathe the roof, but don't nail off the plywood in the area of the dormer. After you've snapped the lines establishing the location of the valley, adjust the depth on your circular saw and cut out the plywood just inside the chalk lines.

The best way to lay out the dormer common rafters is to draw the elevation of the dormer gable full scale on the plywood subfloor (for more on rafter layout, see pp. 76-81 and 82-83). Then you can establish the particulars of the cornice construction at the same time, and simply transfer the cutting angles to your rafter stock with a T-bevel.

Cut out four common rafters and use them to prop up a temporary ridge. With a straightedge, extend the line of the ridge over to the main roof. This will give you the location of the inboard end of the ridge and a point from which to measure the length of the ridge. If the ridge lands on a cripple rafter, its inboard cut will be at the same angle as the level cut of a main-roof common. If the dormer ridge falls between

From *Fine Homebuilding* magazine (August 1987) 41:60-65

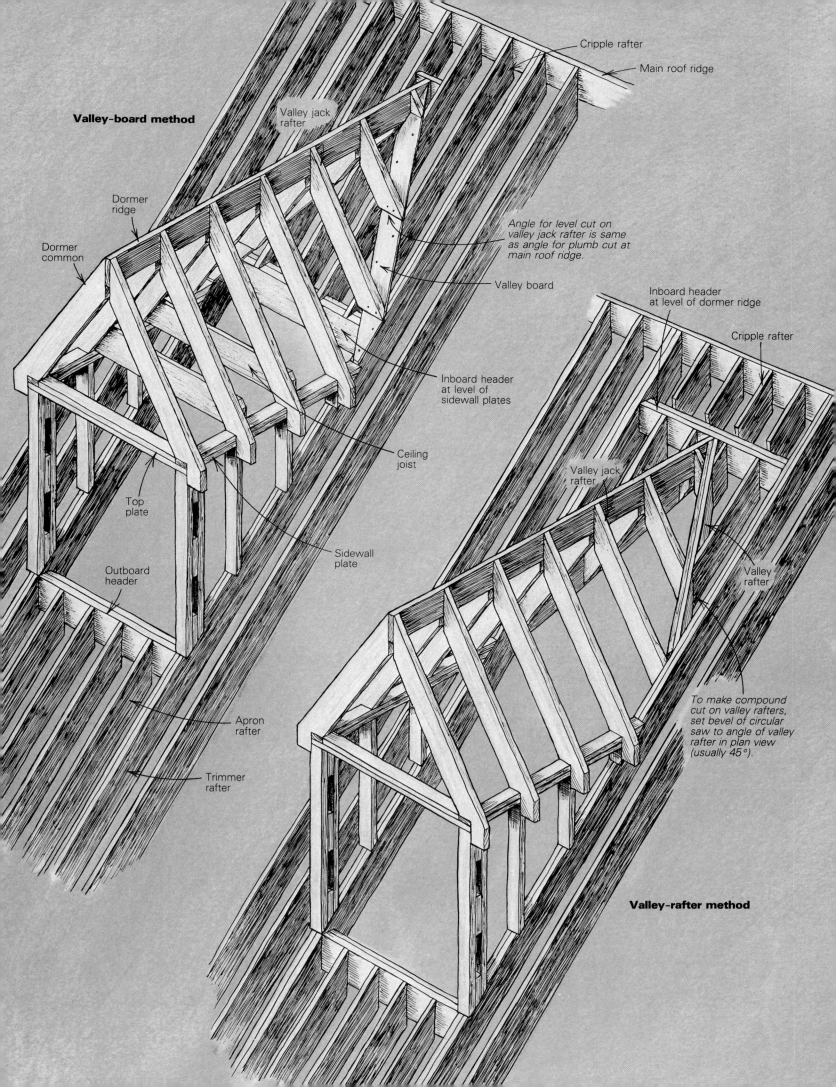

Valley-board method

Cripple rafter

Main roof ridge

Valley jack rafter

Angle for level cut on valley jack rafter is same as angle for plumb cut at main roof ridge.

Dormer ridge

Dormer common

Valley board

Inboard header at level of sidewall plates

Top plate

Ceiling joist

Outboard header

Sidewall plate

Apron rafter

Trimmer rafter

Inboard header at level of dormer ridge

Cripple rafter

Valley jack rafter

Valley rafter

To make compound cut on valley rafters, set bevel of circular saw to angle of valley rafter in plan view (usually 45°).

Valley-rafter method

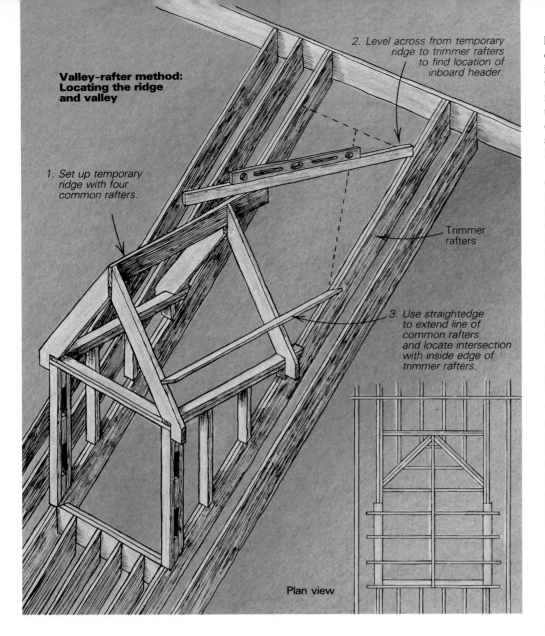

**Valley-rafter method:
Locating the ridge
and valley**

1. Set up temporary
ridge with four
common rafters.

2. Level across from temporary
ridge to trimmer rafters
to find location of
inboard header.

Trimmer
rafters

3. Use straightedge
to extend line of
common rafters
and locate intersection
with inside edge of
trimmer rafters.

Plan view

rafters, you will either have to install an extra cripple, or nail a block between two neighboring ones. In the latter case, the ridge would be cut square to fit against the block, as shown in the drawing at left on the previous page.

Cut the permanent ridge and nail it in place, along with the four common rafters, one pair at the gable end and the other pair at the inboard end of the plate. Leave the rest of the commons out until you've finished cutting and installing the valley jack rafters (the commons will just be in the way otherwise).

Placing a straightedge across the common rafters near the eaves, project the dormer roof plane onto the main roof frame or sheathing. Mark the point of intersection with a pencil. The exact location of this mark isn't critical, as long as it's in the dormer roof plane toward the bottom of the proposed valley. Snap a chalkline over the tops of the rafters from where the upper corner of the dormer ridge strikes the main roof down to your pencil mark.

If the layout of the dormer valley jacks is coordinated with the main-roof cripple rafters, each jack can bear directly over a cripple. In this case, the valley board can be 1x stock, or eliminated altogether. As long as you have direct

bearing on the cripple rafters, you can nail the jacks directly on the main-roof sheathing. If the bottoms of the valley jacks don't line up with the cripples, use 2x stock for the valley board to distribute the load. With a T-bevel, measure the angles for the ends of the valley board, which will be nailed to the main roof.

Since the valley board has thickness, it must be nailed back from the chalkline a bit so that its top outside corner will fall in the dormer roof plane. To determine the offset for the valley board, first take a scrap of 2x4 and put the jack-rafter seat cut on one end. To make this compound angle cut, lay out the level cut used for the dormer common on the face of the 2x4, and cut with the skillsaw set to the plumb-cut angle of the main-roof common rafter.

Tack the 2x4 to the valley board as if it were a valley jack and hold the valley board parallel to the chalkline on the main roof. Now extend a straightedge along the top of the 2x4 down to the chalkline. This will tell you how far to offset the valley board from the chalkline.

Lay out the jacks along the valley board by measuring 16-in. increments off the dormer common rafters closest to the main roof. As you measure, hold the tape or folding rule more or

less parallel to the dormer ridge, and perpendicular to the commons. Lay off the same spacing along the dormer ridge to correspond with the marks on the valley. Now you can directly measure the lengths of the jack rafters, from the uppermost point of the plumb cut on the top end to the toe of the seat cut on the downhill side. This is the longest dimension of the jack. Draw up a list of the lengths.

For each pair of jacks, you need to make only one compound-angle cut, or cheek cut; making the cheek cut on the end of one piece leaves the cheek cut for the opposing jack on the off-cut. Start with a board that is more than twice as long as the rafter you're cutting. After making the plumb cut on one end, measure off the rafter length and mark a level cut on the face of the rafter. Now tilt the circular saw to the plumb-cut angle of the main roof, and make the cut. It saves time to hook up two circular saws—one tilted for the cheek cuts, and one set square for the 90° cuts.

After installing the jacks, put in the remaining common rafters. Then install the ceiling joists and sheathing.

Valley-rafter method—Start by cutting out four common rafters and setting them up with a temporary ridge, just as you would for the valley board method. Extend a level line across from the ridge to the trimmer rafters (drawing, left). Nail the inboard header between the trimmers at this point, and install the main-roof cripple rafters above the header. Just be sure that their top edges are in line with the main-roof rafters.

Now measure for the dormer ridge, and cut and install it along with the four commons. Here, too, it's best to leave out the rest of the commons while you're working on the valleys.

The top of the valley is determined by the intersection of the dormer ridge and the main roof. It's a common mistake to think that the bottom of the valley will be where the sidewall plate strikes the trimmer. I've seen more than one textbook that shows it this way. In fact, that point will end up well below the surface of the dormer roof.

To locate the bottom of the valley, lay a straightedge across the two common rafters and project the line of the dormer roof onto the main roof. You have to find the point where the dormer roof plane intersects the inside edge of the trimmer rafter. This is where the centerline, along the top edge of the valley rafter, will meet the trimmer (drawing, left).

Although in terms of strength, dormer rafters usually need be no wider than 4 in., it's necessary to use stock the same width as the main-roof rafters if you want the dormer and main-roof ceilings to come together in a neat corner. This gives a cleaner look to the interior finish.

The rise of the valley rafter is the same as for the commons (think about it). But the run, on the other hand, is longer, just as it is for a hip rafter. If the pitch of the dormer roof is the same as the main roof (both are 7-in-12 in our drawing), and since they intersect each other at a right angle, the valley between them runs at a 45° angle to the common rafters in plan. This means the valley rafter runs 17 in. for every

12 in. of a common. Therefore, the pitch of the valley rafter is 7-in-17. Using these numbers (7 and 17) on a framing square, you can lay out the plumb-cut angles on the valley rafter. Then cut along the line where the 7 is.

Backing—Now we must determine the width of the valley rafter, and the bevel to be used for "backing" its top and bottom edges. Backing is the process of beveling a hip or valley in the same planes as the adjoining roofs, so that its thickness will not interfere with the sheathing. Hip rafters are usually dropped rather than backed. For the dormer valley rafter, you can skip the backing bevels on the top edge if you wish, by raising up the lower ends of the valley jacks a bit. This will cause the projection of their top edges to strike the centerline of the valley. If the dormer has a cathedral ceiling, however, backing is at least advisable on the bottom edge of the valley rafter. The attic and dormer ceilings will come to an outside corner here (photo next page), and the backing bevels will provide good seating for the drywall and sound nailing for the corner bead.

Backing hips and valleys also clarifies things when lining up the different members in the frame. Since a circular saw will handle the beveling without much trouble, I've found backing to be less bother than the guesswork involved in not backing, especially on tricky roofs.

To simplify the process, I often double the valley and hip rafters, even when doing so is not necessary for strength. This allows me to rip one bevel on each piece before assembly. When the two pieces are nailed up, the opposing bevels form either a concave V-trough or a convex ridge, which are the ideal forms for valley and hip rafters respectively.

Normally when you frame a roof you're concerned with lining up only the top surface of the framing members. The bottom doesn't matter because it's usually an unfinished attic. But when framing a dormer with a cathedral ceiling both the tops and bottoms of the rafters have to line up. The critical dimension is the vertical depth of the rafters measured along a plumb line. But since the valley rafter is rising at a shallower pitch than the commons, you need to cut it from wider stock than you used for your commons (drawing, top right).

To calculate the width you need for your valley rafter, first measure the vertical depth of a common rafter (the length of the plumb cut at the ridge is the easiest place to find this dimension). Then, using a framing square held at 7-on-17, draw lines that represent the pitch and plumb cut of the valley rafter on a sheet of plywood. Measure along the plumb-cut line the vertical depth of the common rafter and strike a line through this point, parallel to the pitch line. The distance between these lines is the width of your valley stock.

To calculate the backing angles for the valley rafter, start by laying the framing square on the face of the rafter stock, as in the drawing at middle right, with the tongue on 7 and the blade on 17 (the pitch of the valley). Now measure along the blade half the thickness of the valley stock, and mark this point on the rafter. Since I

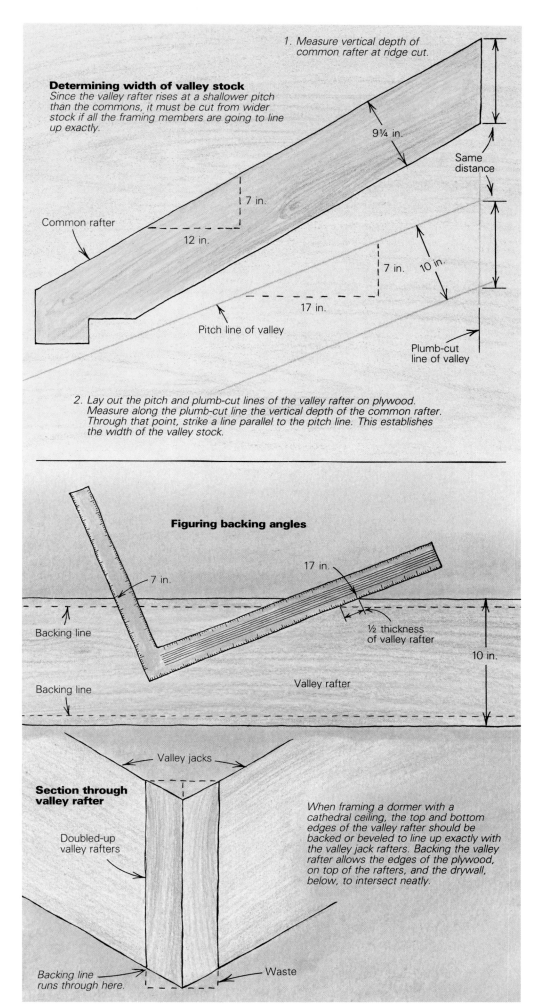

1. Measure vertical depth of common rafter at ridge cut.

Determining width of valley stock
Since the valley rafter rises at a shallower pitch than the commons, it must be cut from wider stock if all the framing members are going to line up exactly.

9¼ in.

Same distance

Common rafter

7 in.

12 in.

7 in.   10 in.

17 in.

Pitch line of valley

Plumb-cut line of valley

2. Lay out the pitch and plumb-cut lines of the valley rafter on plywood. Measure along the plumb-cut line the vertical depth of the common rafter. Through that point, strike a line parallel to the pitch line. This establishes the width of the valley stock.

Figuring backing angles

7 in.

17 in.

Backing line

½ thickness of valley rafter

10 in.

Backing line

Valley rafter

Valley jacks

Section through valley rafter

Doubled-up valley rafters

When framing a dormer with a cathedral ceiling, the top and bottom edges of the valley rafter should be backed or beveled to line up exactly with the valley jack rafters. Backing the valley rafter allows the edges of the plywood, on top of the rafters, and the drywall, below, to intersect neatly.

Backing line runs through here.

Waste

Although this dormer has a hip roof, it was framed with the valley-rafter method and illustrates the clean, geometric ceiling lines that make the complexity of this framing worthwhile.

use a separate 2x for each side of the valley, half the valley stock is 1½ in., but normally it's ¾ in. Strike a line through this point, parallel to the edge of the rafter. This is called the backing line, and the angle between it and the centerline on top of the valley rafter is the backing angle. (This rule holds only for "regular" hips and valleys, meaning those formed by roofs of equal pitch that join at right angles.)

Edge bevel and shortening adjustment— As we have already seen, when equal-pitch roofs intersect at right angles, the resulting run of the valley (plan view) lies at 45° to both ridges (the main-roof header in this case is acting like a ridge). Therefore, making a cheek cut with the circular saw tilted to this angle will produce the correct bevel on the edge of the valley to fit the ridge.

Before it will fit though, the theoretical length of the valley must be shortened to allow for the thickness of the dormer ridge. Since the valley rafter is doubled, we will make the shortening adjustment on each half separately.

In order to fit against the ridge and inboard header, the valley rafter on the main-roof side of the valley gets two cheek cuts (see plan view, drawing at right, p. 114). First, make the cut that will fit against the header through the unadjusted length. No shortening adjustment is made here, because the theoretical layout line coincides with the face (not the centerline) of the header.

For the second cheek cut, measure back from the long point of the first cut, along a horizontal line, one-half the 45° thickness of the ridge (1¹⁄₁₆ in. for 1½-in. thick stock). Mark a plumb

line through this point and make the cut with the circular saw still set at 45°, beveling in the opposite direction to the first cut. If you wanted, you could make the double cheek cut here symmetrical by bringing the header out more to begin with. This would place the theoretical layout line down the middle of the outboard half of the header. But as a result, the header would have to be beveled where its corner protruded above the main-roof surface.

The ridge cut for the other half of the valley rafter simply gets shortened one-half the 45° thickness of the ridge. But if you put the first board in place, it will give you the point from which to measure the actual length of the second valley rafter board.

All these cuts are laid out using the same plumb-cut angle on the face. Shortening adjustments are always made horizontally—that is, perpendicular to the plumb cut.

Jacks— A gable dormer may be so small that it doesn't require any valley jack rafters. If it does require them, lay out the position of the jacks on the top edge of the valley rafter just as you would for the valley board. In this instance, however, the bottom end of the jack will be marked with a plumb cut on its face, instead of a level cut, and the circular saw will be set to 45°, instead of the main-roof plumb-cut angle. The top end of the jack will be cut the same as before—use the plumb cut of the dormer common, with the circular saw set square. □

Scott McBride is a carpenter and contractor in Irvington, N. Y.

Crown molding is often used to trim gable dormers. Its curves and shadows add an elegant flourish to the dormer's overall composition. There are two basic applications. The first has one piece of crown molding along the eaves mitered to another along the rake, with very little overhang on either the eaves or the rake (figure 1, facing page).

The second application, which I call pedimented, continues the crown molding horizontally across the front of the gable, along with the rest of the cornice trim (frieze, bed, soffit and fascia). This forms the base of the pediment (figure 2). A sloped and flashed water table caps the cornice across the front to keep out rainwater. Two additional pieces of crown molding are run along the rake and die into the water table.

This type of pediment is traditional over entries and windows, as well as dormers, in the classically derived styles. Sometimes the cornice turns the corner and extends only a foot or so onto the gable end—just enough to provide a neat terminus for the rake crown. This is called a cornice return, or sometimes a boxed return. Pedimented gables are a lot of work, but their sharp appearance usually justifies the effort.

Turning the crown molding directly up the rake might seem the simpler of the two applications, but it isn't because of the miter. The crown molding changes planes as it turns upward, so if it's set in the usual way along the eaves, it won't match up with the molding along the rake. There are two ways to deal with this problem.

The first is simply to tilt the crown molding down at the eaves, so that its top edge lies in the roof plane, as shown in figure 3. Tapered blocking nailed to the fascia will give proper bearing for the back of the eaves-crown molding. It can then be mitered to the rake crown with a regular 45° miter.

You can get away with this approach on most low-pitched roofs. As the roof pitch of a dormer increases, however, this method forces the face of the crown molding to lie flat, contrary to its original intent.

The "correct" way to handle the situation is to have milled a molding with a slightly different profile just for the rake return, called a "raking molding" (not to be confused with the standard "rake molding" sold at lumberyards). This modified crown molding will have the same horizontal depth, front to back, as the crown molding at the eaves, but its vertical height will be stretched out a bit, depending on the pitch. The development of the rake crown's profile is shown in figure 4. The depth of the eaves crown at different points, front to back, is transferred to the rake crown unchanged, using arcs and parallel lines. Notice how the *diagonal* distances between points on the *face* of the eaves crown have become *vertical* distances between points on the *back* of the rake crown. When further projected onto the face of the rake crown, the diagonal distances between corresponding points increase. Note that the run of the eaves crown, shown in figure 4 as

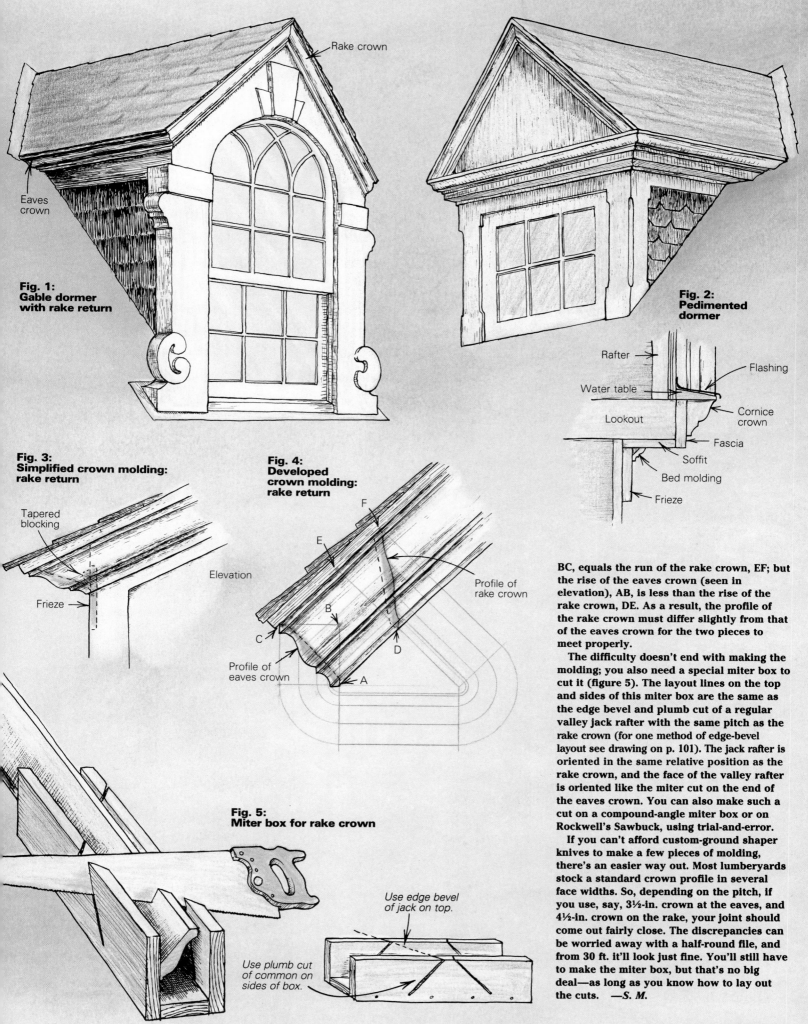

**Fig. 1:
Gable dormer
with rake return**

Rake crown

Eaves
crown

**Fig. 2:
Pedimented
dormer**

Rafter

Flashing

Water table

Cornice
crown

Lookout

Fascia

Soffit

Bed molding

Frieze

**Fig. 3:
Simplified crown molding:
rake return**

Tapered
blocking

Frieze

**Fig. 4:
Developed
crown molding:
rake return**

Elevation

F

E

B

C

D

A

Profile of
rake crown

Profile of
eaves crown

**Fig. 5:
Miter box for rake crown**

Use edge bevel
of jack on top.

Use plumb cut
of common on
sides of box.

BC, equals the run of the rake crown, EF; but
the rise of the eaves crown (seen in
elevation), AB, is less than the rise of the
rake crown, DE. As a result, the profile of
the rake crown must differ slightly from that
of the eaves crown for the two pieces to
meet properly.

The difficulty doesn't end with making the
molding; you also need a special miter box to
cut it (figure 5). The layout lines on the top
and sides of this miter box are the same as
the edge bevel and plumb cut of a regular
valley jack rafter with the same pitch as the
rake crown (for one method of edge-bevel
layout see drawing on p. 101). The jack rafter is
oriented in the same relative position as the
rake crown, and the face of the valley rafter
is oriented like the miter cut on the end of
the eaves crown. You can also make such a
cut on a compound-angle miter box or on
Rockwell's Sawbuck, using trial-and-error.

If you can't afford custom-ground shaper
knives to make a few pieces of molding,
there's an easier way out. Most lumberyards
stock a standard crown profile in several
face widths. So, depending on the pitch, if
you use, say, 3½-in. crown at the eaves, and
4½-in. crown on the rake, your joint should
come out fairly close. The discrepancies can
be worried away with a half-round file, and
from 30 ft. it'll look just fine. You'll still have
to make the miter box, but that's no big
deal—as long as you know how to lay out
the cuts.   —*S. M.*

Decking and Sheathing

How one production carpenter puts the skin on wood-frame houses

by Jud Peake

The skin that knits together a building's bones and provides a base for the finish materials is the sheathing, and in contemporary wood-frame construction that usually means plywood. Sheathing a building isn't a fussy or complicated operation. Thousands of square feet of plywood have to be secured with tens of thousands of nails, and a lot of odd-sized pieces of plywood (specials) have to be cut to fit the parts of the building that don't conform to a 4-ft. by 8-ft. module. Fortunately, a skillsaw makes short work of the cuts, a nail gun takes the pressure off your forearm and at 32 sq. ft. per sheet, plywood covers big chunks of a building all at once. Given the simplicity of the task and the capabilities of the tools, clearly the thing to learn about sheathing is how to do it fast.

Subfloor—I use T&G plywood almost exclusively for subfloors because it eliminates the need for blocking under the long edges of the plywood, and so cuts joisting time in half. A plywood subfloor should start on a side of a building that is perpendicular to the joists. It's nice to have a straight side to begin the plywood; if the building has projections for bays and rooms, I begin with the side that has the fewest jogs. I snap a chalkline across the joists 4 ft. in from the edge of the building and lay down a bead of construction adhesive on the tops of the joists that will fall under the first course. I align the plywood so that its groove side butts the chalkline, facing the bulk of the building, and I make sure I've got about a 1/16-in. gap between the ends of the neighboring plywood panels to allow for expansion. Then I nail off this first course, but I don't put any nails within 6 in. or so of the grooves. This allows me some wiggle room to install the next course of plywood.

If you are working on a second floor and a forklift is available, now is a good time to land a stack of plywood on the deck. Let the stack hang a little over the edge of the building so that you'll have room to work the next course into the first course of grooves. As an alternative you can land your plywood stack on the side of the building that you are working toward.

Sheathers generally work in pairs—one carpenter spreads the plywood and nails it off, while the other one does any necessary cutting, applies the construction adhesive to the joists and fits the T&G edges of the plywood together with a bump-stick. The bump-stick eliminates the need for another carpenter and a sledgehammer. To make a bump-stick, simply lap and nail two 2x4s together in the shape of a T. A 4-ft. crosspiece will do, and the handle should be about 5 ft. long with about a 20° bevel on the end that is attached to the crosspiece. To use the bump-stick, stand with your feet on the joint between the plywood sheets to align the tongues and grooves, then pull and bump the plywood together with the stick (photo below).

Four sheets of plywood should never come together at one corner, so the second course of subfloor should be staggered by starting it with a half-sheet of plywood. This ensures that the end joints of two sheets are aligned at the corners by the tongue or groove of the neighboring sheet of plywood, and strengthens the overall floor diaphragm by reducing long seams between adjoining materials. If your layout allows you to use the same size pieces to start staggered rows, precut them on the stack. To speed things up, I use a T-square designed for marking drywall to lay out the cutlines on plywood sheathing.

To get the best value for your money, get the big glue gun and the 32-oz. tubes of construction adhesive. Although the American Plywood Association recommends putting adhesive in the T&G joint, I've never met anyone who has done it with success. The proper size bead of adhesive for the joist tops (about 1/4 in.) is far too much glue to fit in the T&G joint.

Along the leading edge of each course of plywood mark the centerline of each joist as a guide to help the nailer. If you're using a nail gun and are new to production sheathing, a 4-ft. 2x4 can be a fence for positioning the nails. (With practice you won't need it.) Line up the 2x4 with the joist marks on the plywood, then stand on the 2x while you use its edge to guide the tip of the gun. For subfloors, the perimeter of the plywood sheets needs nails 6 in. o. c.; the field nails fall at 10-in. intervals. Use 6d nails to secure floors less than 5/8 in. thick. Floors that are 5/8 in. or 3/4 in. thick need 8d nails.

The subfloor should be nailed off while the glue is still wet, but don't nail off the edges around the perimeter of the building until the very end because nail heads here may be in the line of cut. Instead, let the plywood overhang the framing, snap a chalkline to indicate the line of the floor, cut off the overhang with one pass of the saw and add the missing nails. If you have

The bump-stick is a T-shaped tool made of framing lumber that can help when you're installing T&G plywood. Align the tongue on the workpiece with the grooves in the installed sheets by standing on them. Then pull on the bump-stick to drive the sheets together.

From *Fine Homebuilding* magazine (February 1987) 37:66-69

any interior stairwells, be sure to leave sufficient subfloor overhang for the top nosing.

Floor joists often lap one another in the center of a floor. Where this happens, you should nail a block to the joist to pick up the unsupported edge of the plywood (middle photo below). Also, when a sheet of plywood lies over the lapped joints, you should make V-shaped marks along the edge of the plywood to denote the centers of the joists and the direction they run to make sure your nails end up in the right place.

Plywood sheathing, be it subfloor, roof or wall sheathing, should have at least ½ in. of bearing on the framing members. But sometimes a bowed joist will leave a sheet a bit shy of the necessary purchase. When this happens, push the joist into alignment and angle an 8d toenail upward to grab the ply (drawing A, below). A layout mistake may place a joist just beyond the edge of the plywood. If this happens, it's usually best to cut the plywood back to the center of the previous joist. These fixes also apply to bowed or misplaced rafters or studs.

If the kitchen and bathroom plumbing have been roughed in, you'll have to contend with all the pipes that penetrate the subfloor in these areas. This is a good time to work with your helper—one person calling out the locations of the pipes while the other marks them down on the plywood (photo below right). Just measure the two coordinates for the centerline of a pipe—this cuts in half the numbers you have to remember. I drill holes for pipes up to 1 in. in diameter. I make the holes ¼ in. oversize to allow for maneuvering the plywood into place. For larger holes I use a skillsaw or a jigsaw.

Wall sheathing—Plywood's tremendous shear values make it especially good for rough sheathing on walls. And just as T&G plywood elimi-nates the need for subfloor edge blocking, oversize plywood can save you the trouble of blocking the walls. You can get square-edged sheets of CDX that are 4 ft. wide, and 9 ft. or 10 ft. long. The extra material gives you plenty of overlap to pick up mudsill, joists and top plates on a building with 8-ft. ceilings.

If it is to be covered with siding, ⅜-in. CDX is the typical sheathing. Secure it with 6d nails, 6 in. o. c. around panel perimeters and 12 in. o. c. in the field. In earthquake-prone areas, engineers will often specify nails at closer intervals.

Although it isn't required by code, I like to use ⅜-in. CDX behind stucco, especially in any area subject to contact with baseballs or bicycles. If the sheathing is also going to serve as a nailing surface for sidewall shingles, it should be at least ½ in. thick, secured with 8d nails.

To allow for moisture-induced expansion, you should leave ⅛-in. gaps between the plywood panels, especially those parallel to the face grain. This presents a problem with conventional stud layout. Every sixth sheet will completely miss the stud. I deal with this by ripping ⅜ in. off every third or fourth sheet.

Typically, wall-sheathing end joints will fall on the rim joists of multi-story buildings. As you sheathe the upper stories, snap a chalkline onto the rim joists ½ in. above the first-floor wall sheathing. Then tack nails into the line to give yourself a ledge to rest your panels on. This gap allows for the shrinkage of plates and joists.

Sheathe all the walls without regard for doors or windows. Come back when the building is completely sheathed and cut out the window and door openings with a chainsaw or a Sawzall from inside. Be sure to keep nails away from the rough openings until you've made your cutouts.

For me, the scariest moment that I commonly face in construction is raising pre-sheathed walls, two or three floors up, on a windy day. Under these conditions walls want to become hang gliders, taking reluctant carpenters for short but memorable rides. Consequently, I much prefer to affix plywood to plumbed and aligned stud walls from a scaffold. But sometimes walls must be sheathed before they are raised—for instance, a blind wall hard against a neighboring house on a zero-setback lot.

When I have to raise a pre-sheathed wall, I snap a line on the subfloor to register the inside edge of the plate. Then before raising the wall I toenail the plate to the subfloor with 16d nails on 4-ft. centers. As the wall is raised, the nails work as a hinge to keep the plate from moving beyond the layout line. If I need to bring the wall back a bit, I dig my hammer claws into a stud for purchase.

Although pre-sheathed walls are squared before raising, they rarely stand perfectly plumb. Since you have to get on a ladder anyway to nail off the plywood lap from the butt-wall to the by-wall ("Stud-Wall Framing," pp. 41-49), ignore the corner framing members on both walls until you've checked the walls for plumb. This strategy makes it easier to adjust the walls, and all the corner framing and sheathing can be nailed off at the same time.

Roofs—If a roof has eaves, the sheathing begins with the starter boards. These are usually 1x6 or 1x8 shiplap boards that show on the underside of the eaves. On tract work the starter boards are often the responsibility of the same two-man crew that installs the barge rafters and fascia, while another crew nails down the rest of the roof sheathing.

Begin the starter-board installation by checking the alignment of the rafter tails with a string line. If any of them are out (a common problem

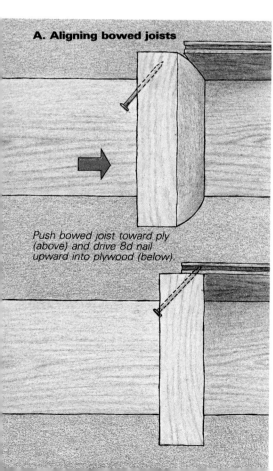

A. Aligning bowed joists

Push bowed joist toward ply (above) and drive 8d nail upward into plywood (below).

When joists are lapped, you have to add a block to pick up the unsupported edge of a piece of subfloor. There's a V-shaped mark on the plywood to the right of the hammer. It tells the person nailing where the centerline of the joist lies, and the direction it runs.

Try to work in pairs when laying out the bathroom subfloor. One calls out pipe centerlines while another marks their locations.

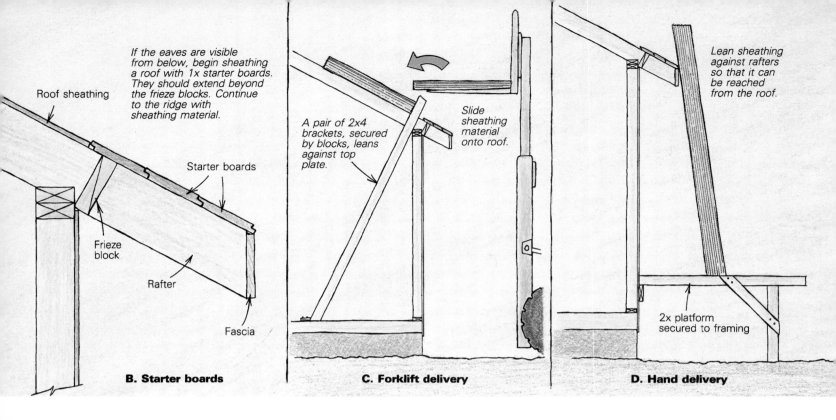

If the eaves are visible from below, begin sheathing a roof with 1x starter boards. They should extend beyond the frieze blocks. Continue to the ridge with sheathing material.

Roof sheathing

Starter boards

Frieze block

Rafter

Fascia

B. Starter boards

A pair of 2x4 brackets, secured by blocks, leans against top plate.

Slide sheathing material onto roof.

C. Forklift delivery

Lean sheathing against rafters so that it can be reached from the roof.

2x platform secured to framing

D. Hand delivery

Cutting two pieces of roof sheathing at once eliminates tedious measuring and unwieldy rooftop cutting setups. Workers below should be warned about falling offcuts.

with roof trusses), snap a chalkline and make the necessary plumb cuts. While you are at it, snap another chalkline to show where the frieze blocks go, and a third to locate the line of the starter boards. The line for the starter boards will be a distance that is a multiple of their widths, and it will fall on the ridge side of the frieze block (drawing B, above left). I begin my starter boards on this chalkline because rafter tails are too awkward to use as a register mark when working from above.

If your supplier can bring along a forklift when he delivers the roof sheathing, build a 2x bracket to cradle the sheathing pile near the eave line (drawing C). Lacking a forklift, build a staging platform that is high enough for a vertical stack of sheathing to be accessible from the roof (drawing D).

If the pitch is flat enough to need a tar-and-gravel roof or some other type of membrane waterproofing, use T&G plywood sheathing. A roof that is steep enough to use shingles doesn't need T&G plywood. T&G sheathing is difficult to install properly on a slope, and it adds nothing to the durability of the finish roofing.

I have occasionally seen flat roofs that have been specified to have ½-in. square-edged sheathing supported at the unblocked edges by ply-clips. Ply-clips are H-shaped aluminum extrusions that fit over adjoining pieces of plywood, linking the two. Because ½-in. plywood isn't available in T&G, the ply-clips are a cost-cutting strategy. Considering the time and effort it takes to install the clips, I say skip the clips and pay for ⅝-in. T&G sheathing.

With square-edge plywood roof sheathing, you can make sheathing cuts similar to skip sheathing (photo left). To note nailing centers, mark the rafter or truss locations on the edge of the plywood as you go. Standard nailing for ½-in. to ¾-in. sheathing is with 8d nails, 6 in. o. c. around the perimeter of panels and at 12-in. intervals in the field.

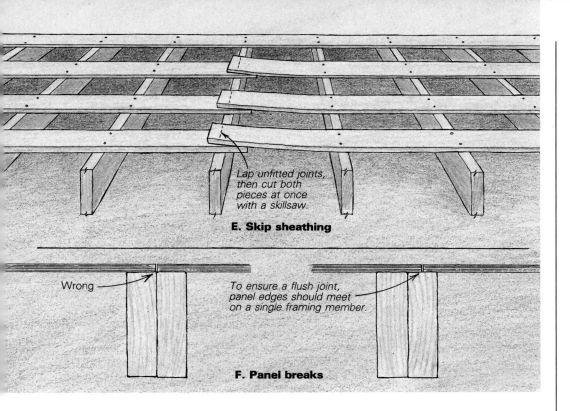

E. Skip sheathing

Lap unfitted joints, then cut both pieces at once with a skillsaw.

Wrong

To ensure a flush joint, panel edges should meet on a single framing member.

F. Panel breaks

Skip sheathing—Wood shingle and shake roofs remain the last refuge of lumber (board) rough sheathing. When I first started as a production carpenter 14 years ago, we used to cover the rafters about halfway up the roof with 1x4 and nail every other one. Then we would pull out the alternating 1x4s, which had served their jobs as spacers, and move them uphill where the process would be repeated. Hence the name skip sheathing. Wood shingles and shakes are applied over skip sheathing because the gaps between the boards allow air circulation to keep the roofing dry.

Nowadays it's more common to see 1x6 skip sheathing laid center to center (with a gauge mark on the hammer) to a measurement that corresponds to the exposure of the shingle, shake or even tile.

Long boards are best for simple gable roofs, but they can be a nuisance on roofs with a lot of hips and valleys. Consequently, lean the long stuff around the house against the eaves where you can reach it from the uncomplicated parts of the roof, and save the short pieces for the complex areas.

As you lay up the 1xs, using either the skip sheathing method or a gauge-mark on your hammer, don't stop to cut the material to length. Instead, let the pieces lap over one another for the time being. Make sure each 1x spans at least two rafter bays, and you can lap up to three butt joints in a row on the same rafter (drawing E, above).

Use 8d nails for 1x skip sheathing—two at each rafter for 1x4s and 1x6s; 1x8s and up get three. Keep the nails out of the way of the future line of cut. After you get the whole roof laid up, grab your saw, set its depth to cut two thicknesses of skip sheathing, and do all the fitting by cutting through both the over and underlapped piece at the same time. Sheathe solid the last three courses at the ridge.

On roofs above a 6-in-12 pitch skip sheathing is nice because it makes a ladder to work on. A plywood roof at 6-in-12 is only deceptively comfortable—a little sawdust can start you on a slide that will have you using your hammer claws, fingernails and newfound ability at rapid prayer. Roof jacks or rows of 2x4s tacked 4 ft. apart across the roof can help the sheather as much as the shingler.

Engineering—Except in engineered cases (like preframed roof panels), always run the face grain of the plywood perpendicular to the joists or rafters. Most span tables for plywood and other coverings assume that the sheathing will be at least two spans long. For shear walls engineers often specify a minimum size panel. Panel ends and edges should break on a single member—not align with the gap between two pieces nailed together (drawing F). This helps ensure a flush joint and continuity of shear strength.

Unlike floors and roofs, you don't need to stagger the joints in wall sheathing unless specified by the engineer, and the face grain of the plywood can be run parallel to the studs. It is however, especially important to nail wall sheathing properly because it is usually very thin. Nails should be no closer than ⅜ in. to the edge of the sheet, and they should not tear through any of the laminations. I've actually seen some engineers specify hand nailing to avoid this problem. To get a good job without going to such a ridiculous extreme you can shim out the nose of the nail gun to leave the nail head out about ¼ in. Then go back and finish off the nails with a hammer.

A 14-gauge staple will provide almost identical shear strength and greater resistance to withdrawal than an 8d nail, while doing much less damage to the plywood. I don't understand why engineers don't specify them more often.   □

Jud Peake is a contractor from Oakland, Calif., and a member of Carpenter's Local #36.

Types of sheathing

Plywood is graded by the quality of the face veneers—A through D. An A rating is the most defect-free plywood grade, and D has the most defects. The grade is marked out on the back of each sheet with a stamp that notes the grade of the face veneer, followed by the grade of the back side. For example, a sheet of AD plywood has had all its voids patched with wood plugs on one side (the A side), and then it has been sanded smooth. The back side, or D side, is unsanded and has voids from splinters and knotholes up to 2½ in. in diameter.

Plywood can be made from more than 70 species of wood, all of which have different strength ratings. The ratings are divided into five groups, with the strongest being the group 1 species. Plywood made only from the group 1 species will bear the label "Structural I." It is suitable for use in highly demanding applications such as box-beams or structural diaphragms. Structural II plywood is composed of woods from group 1, 2 or 3. Plywoods that don't have these labels can be made of any of the five groups.

The veneers that make up a piece of plywood can be bonded together with two kinds of glue. Interior plywood is made with water-resistant glue; exterior plywood is made with waterproof glue. Most sheathing plywood is made with exterior glue to hold up while exposed to the weather during construction. The grade of sheathing common to most residential projects, CDX, is actually an interior grade of plywood because exterior grades don't include a D face. The X in the label means the plywood was made with exterior glue.

Plywood that is suitable for sheathing will have a label that notes the span index. This tells you the allowable distances between rafters or joists for the grade and thickness of the plywood. For instance, the stamp 32/16 means that the plywood is acceptable (under most codes) for sheathing roofs with rafters no more than 32 in. o. c., or floors with joists a maximum of 16 in. o. c. The stamp 24/0 means the plywood is good for roofs with rafters no more than 24 in. o. c. but unacceptable for subfloors.

Typically, a house with joists 16 in. o. c. will have a ⅝-in. thick CDX subfloor, and joists 24 in. o. c. will have a ¾-in. thick CDX subfloor. If the floor finish is to be vinyl, the subfloor can be CDX that has been plugged and touch-sanded. But in practice, most builders I know add a ⅜-in. thick underlayment for vinyl floors because it is smoother and less likely to be damaged during construction.

While plywood has been the common rough sheathing on homes built since World War II, two new types of sheathing have recently come on the scene. One is oriented strandboard; the other is oriented waferboard. Both are made by bonding together with phenolic resins flakes or chips of wood from trees that are unsuitable for plywood. Both types come in T&G, and in the standard plywood dimensions. They are manufactured to standards set by the American Plywood Association, with span indexes printed on each panel. If you are close to a mill that produces oriented strandboard or waferboard, chances are they will cost up to 15% less than comparable sheets of plywood. For more information on panel sheathing products such as plywood and waferboard, write the American Plywood Association, P.O. Box 11700, Tacoma, Wash. 98411.   —*J. P.*

Deck Design

A guide to the basics of deck construction

by Scott Grove

A properly built deck should last a lifetime. But for this to be possible, you must constantly think about nature's elements as you design and build it. If you don't, trapped moisture can promote bacterial degrade that will slowly eat your deck away.

Planning the deck—A deck is an intermediate space between the controlled environment of a house and the raw elements outdoors. Since a deck can expand the living area of the house and serve as an entry, it's important to consider traffic patterns in your planning. Avoid paths that cross through activity areas, and arrange for them to be as direct as possible. A path improperly located can isolate small areas and render them nearly useless.

A deck can accentuate the good features of an area and minimize the bad ones. It can conceal a fuel tank or snuggle around a tree. Decks are great for hiding ugly foundations, service meters or old concrete patios. Let these existing elements influence your deck design and they'll make your job easier. The space under a deck can be used to store firewood, too.

Safety is an important consideration when de-signing a deck. For instance, a landing in front of a door needs plenty of room to allow the door to open with at least one person on the landing. A low walkway that may be just fine without a railing in the summer can be most dangerous in the winter, when snow conceals its edges. Define these edges and all corners, using posts, trees, bushes, rocks or any other visual device on or off the deck that will help make the feature more obvious.

Designing a deck can seem complex if you've never built one before, so beginners should make a detailed drawing of the entire layout, board by board. Once the design is on paper, it's fairly easy to compile a list of materials. Planning the layout and orientation of a deck is at least as important as building it.

Estimating costs—We've been building decks in New York State for seven years, and we use the following figures for rough estimates of the materials and labor needed to build a deck: $8 per sq. ft. for decking (including the framing and footings), $7 per sq. ft. for stairs, $10 per lin. ft. for simple railings, $15 per lin. ft. for bevel-cut railings and $20 per lin. ft. for benches. These figures reflect our company's wage scale, construction speed and craftsmanship, and if a deck design is particularly unusual we'll adjust the figures upward. For those who work alone or with minimal help, the following materials-only estimate, based on prices for #1 pressure-treated lumber in New York State, will help determine approximate costs: $3.00 per sq. ft. for the decking lumber, $4.50 per sq. ft. for stairs, $2.25 per lin. ft. for railings, and $8.25 per lin. ft. for benches. The type of construction you use, and the level of detail you include, will have a significant effect on the expense of your deck.

Choosing lumber—Water is the worst enemy of woodwork, and this fact should be foremost in your mind as you select lumber for your deck. Remember that water does the most damage when it rests undisturbed on or in the wood, especially in places that are slow to dry out. Warping is the number-one problem with decks, and water contributes to the problem. Checks channel water inside a board to accelerate the decay process, and so when we're building a deck we routinely cut back boards with serious end checks. We allow for these cuts when we design a deck by making sure that our plans call for material about 6 in. shorter than standard lumber lengths.

The amount of moisture within new lumber determines how much it will shrink. In wood that's continually exposed to the weather, shrinkage can be considerable. Try to buy kiln-dried lumber, even if this means purchasing from a supplier other than the one you usually use. If dry wood is not available, or if its added cost is not in your budget, at least make sure that the moisture content is consistent throughout your selection. You may not be able to prevent shrinkage, but if you plan for it ahead of time the deck will look better because the gaps between the boards will look uniform. Different-size gaps will make the work look sloppy.

The grade and species of lumber you select will directly affect the longevity of your deck. In parts of the West, decks are frequently built from cedar or redwood. These species are readily available and quite resistant to decay. But in the eastern part of the country, pressure-treated lumber is used most often because it withstands our harsh climate, and is generally more available and less expensive than cedar or redwood.

There are two grades of pressure-treated lumber suitable for decks. We strongly recommend using #1 yellow pine, particularly for the rail-

Decks ease the transition between the house and the landscape, and also serve as an entry. Decks should be functional, durable, well proportioned and attractive. Properly designed so they won't trap water, they will withstand the destructive forces of the weather.

From *Fine Homebuilding* magazine (October 1985) 29:42-46

ings, benches and decking. The quality of #1 pressure-treated lumber is fairly consistent, and the material is easier to work than #2 grade. The #2 grade has a greater number of open knots, and these weaken the boards and encourage water to accumulate. Large knots that span more than half the width of a board are very dangerous in either grade, since the pressure of a footstep or rough handling during construction will sometimes snap the board in half. One problem with pressure-treated lumber is warpage. It can twist severely, cup and bow if not handled correctly. Keep it covered and out of the sun until you use it.

Pressure-treated lumber often has a greenish color, due to the chemicals it's impregnated with (usually chromated copper arsenate). This tint will weather away into a pleasing light grey in about two years, though the treatment chemicals still protect the wood. Some people want more color to their deck, however, so we recommend a semi-transparent stain. If you wait a year or so before the first application, the wood will have a chance to dry out and will accept a fuller coating, doubling the stain's expected life. We prefer stain to paint because paint traps moisture and requires more maintenance.

When you order the lumber for a deck, include about 10% more than you think you'll need. This will prevent time-consuming trips to the lumberyard if your estimate was slightly off, and allow you to cull out badly warped boards with too many knots.

When the lumber is delivered, remember that moist lawns and delivery trucks are a bad combination. There are better ways to find out where the septic-system drainfield is than to have a truck crush the drain tiles. And since the chemicals in pressure-treated lumber can kill grass, make sure you relocate lumber piles after three days to a different location on the lawn.

Nails—We use only galvanized nails on deck projects, 10d for the decking and 16d for framing. Two types of galvanized nails are available. Electro-plated nails have a smooth finish and take less effort to pound in, but hot-dipped nails, with their rough surface, grip much better and are also more rust-resistant. To save time in laying down decking, we use a pneumatic nailer and resin-coated galvanized nails. The resin coating heats up when the nail penetrates the wood, and then hardens like glue for a firm grip.

If you have problems with lumber splitting as you nail into it, use your hammer to blunt the end of your nails. This way they will puncture the wood instead of piercing and splitting it.

Piers—Like the foundation of a house, the foundation of a deck must transfer loads from the structure to the ground. But unlike the foundations of most houses, deck foundations are not continuous. To support the deck, a system of concrete piers is used. The piers extend from grade level to below the frost line—32 in. to 48 in. in our climate. A pier that does not go below the frost line will eventually heave and push the deck out of level.

The standard method of determining where the piers will go requires string, a collection of stakes, and the application of some basic practical geometry. This method works well on decks with simple rectilinear froms, but complex forms are considerably trickier to deal with. When we are faced with the task of building elaborate forms, we've found a way to locate piers that works quite well, and that allows for design flexibility as the project progresses.

Rather than spend a lot of time and effort to locate all the piers at once, we use a locate-build-locate process. The idea is to define the limits of the deck, locate and then pour perimeter piers. Once this is done we can frame the perimeter of the deck, bracing it in place. After that, we locate the rest of the piers.

The easiest piers to locate are the ones that must be placed at a particular point. If you know, for example, that you want the edge of the deck to change direction about 10 ft. from the house and 14 ft. from the oak tree, dig and pour a pier there. The process is empirical: you build what you know in order to answer questions about what you don't know.

After the concrete has partially cured in about 24 hours (it will take nearly a month to gain most of its strength), we begin framing. This method may seem somewhat backward, but we often find it much easier and more accurate in the long run to dig some piers to support interior spans after the perimeter is established.

You will need a long-handled shovel, a digging bar and a post-hole digger to dig the holes for piers. There are two kinds of manually operated post-hole diggers, and you may end up using both of them on your deck project. A post-hole auger looks and works somewhat like a giant corkscrew; as you turn it into the earth it pulls dirt from the hole. An auger works particularly well in hard ground, but is easily stymied by rocks. A clamshell post-hole digger looks like two long-handled spades hinged together at the ferrule, with the blades opposing each other. The work goes quickly in soft ground, but more slowly in packed or clay soils. The clamshell is less likely to stall when you hit rocks, since it can reach into a hole to remove them, but large rocks can cause problems.

If you have a lot of holes to dig, you can rent a gasoline-powered hole digger. This is basically a power auger, and we prefer the one-person model with a torque bar because it won't take you for a ride when it hits a rock.

A long digging bar comes in handy for loosening dirt and breaking rocks that can't easily be removed from the hole in one piece. It also helps loosen tightly packed soil. This solid, heavy, steel persuader is pointed on one end, and can also be used to pry out rocks.

Rocks are the main problem in digging footings around here, but roots can also be a nuisance since decks are frequently near large trees. Use an old handsaw or sharp ax to cut the roots cleanly, but don't seal the cut ends. A botanist once told me that a root or branch will heal itself, and that tar and other sealants interfere with this process.

The shape of the holes you dig is nearly as important as their depth. They should generally be round, and about 8 in. to 12 in. in diameter. The sides of the hole should be reasonably smooth, and the bottom of the hole should be slightly larger than the top to distribute loads well. If there are any ledges or if the hole narrows at the bottom, the freeze/thaw cycle will lift or tip the pier as much as 12 in. over time. Make sure that the floor of the hole is undisturbed earth, because a layer of soft earth here will allow the pier to sink.

We usually pour concrete directly into the hole, using the sides of the hole to form the pier. You can also use Sonotubes to line the hole. These cardboard tubes, available in various diameters from masonry-supply stores, are especially handy if you want the concrete to extend above grade to form a pier. If you suspend the tube 6 in. above the bottom of the hole when you pour, the concrete will ooze out the bottom to widen the base of the pier and increase its bearing ability. Piers should include #4 reinforcing bar if they extend more than 6 in. above grade.

A good concrete mix for piers is 1:2:3, which means one part portland cement, two parts sand, and three parts gravel (¾-in. or 1-in. gravel will be fine). An alternative to mixing your own concrete is to purchase ready-mix, which is a pre-proportioned cement, gravel and sand mixture that usually comes in 90-lb. bags. The portland cement will sometimes settle to the bottom of ready-mix bags, so its a good idea to dry-mix the contents of each bag before adding water. A wheelbarrow is great for mixing concrete in, but be sure to wash it out afterwards, along with your mixing tools.

Finding level—Building a deck can be an exercise in elementary civil engineering, and many beginners are frustrated by having to find the proper relationship between posts and boards that aren't connected. You can't always use a carpenter's level to do this—how would you check two posts, 15 ft. apart, to see if they are at the same height? We often use a 2-ft. level on a long, straight board to check for level, but other tools can be used as well.

A string level is a small spirit level that hooks onto a length of layout twine. When the twine is pulled taut, a rough estimate of level can be determined by raising or lowering one end of the twine and watching the bubble in the level.

An optical pocket level is something of a cross between a telescope and a transit. Looking through it, you align a small leveling bubble with cross hairs to determine an approximately level visual line.

A water level is an inexpensive and very accurate homemade device used to check the relative heights of widely separated items. It's made from clear plastic tubing filled with water (a few drops of food coloring will make the water easier to see). Because of atmospheric pressure, the water level at one end of the tubing always matches the water level at the opposite end, no matter how many twists and turns the tubing takes. It's particularly useful over long distances, as when you want to compare the heights of ledger and posts.

A transit is a precision instrument used by surveyors, and this is what we use to determine level, plumb and the relative heights of widely distant objects with a high degree of accuracy. The transit is fairly expensive, but if you do a lot of decks, the money is well spent.

The ledger—The ledger is a length of 2x lumber that is attached directly to the house, allowing a portion of the deck to "borrow" the foundation of the house for support. It's usually the first framing member to be installed and should be selected from the straightest stock available, since it serves as a reference point for much of the work to follow. When you install the ledger, don't rely on siding or the foundation to be level, because often they're not.

The top of the ledger supports the decking, and if you think of it as a rim joist, you'll get the idea. If the ledger has to be attached to the house foundation in order for the final deck elevation to be where you want it, you'll have to fasten it with lag bolts and lead expansion shields or some other masonry-anchor system. Masonry nails won't work very well, particularly in poured foundations that have had many years to cure. When using a standard masonry bit in an electric drill to bore holes for the expansion shields in a concrete foundation, use a star drill to break apart any pieces of aggregate you can't drill through. We've found that a roto-hammer speeds this job considerably.

Sometimes the plans will call for the ledger to be fastened above the house foundation, and in this case, 4-in. by ⅜-in. galvanized lag bolts spaced about 24 in. apart and fastened to studs or a rim joist will usually do the job. Slide flashing under the existing siding and over the ledger, if possible, to keep water from seeping behind it. If the ledger is going to be mounted to some sort of concrete patio or walkway, use shims to hold the board away from the concrete to allow the water to pass freely by.

Allow at least 1 in. between door sills and the decking surface to prevent any water from running back off the deck into the house. If you include a step, use the same step rise used elsewhere on the deck for the sake of consistency.

Posts—Posts transfer the loads from the deck structure to the piers. One end of each post is attached to a beam or a joist, and the other end rests on the pier. We usually don't anchor posts to the piers, since the weight of the deck is enough to keep them in position. Though many people embed posts in concrete, we feel this technique can create serious problems if water collects between the post and the concrete. Posts are usually 4x4s, 4x6s or 6x6s. A 6x6 post is generally more than needed for bearing purposes, but it enables us to notch beams or joists into it for added strength.

Beams—Beams are an intermediate structural member, used to support joists. They can be solid lumber, usually 4x6 or 4x8, or they can be built up out of 2x lumber. When you're fabricating built-up beams, a common mistake is to nail the individual 2x material face to face, which allows water to get trapped between these boards. Instead, you should sandwich blocks of ½-in. pressure-treated wood between the boards to create a void for water to run through.

Joists—Joists are the uppermost structural element supporting the decking. They are generally 2x lumber, and should be reasonably straight. When laying joists into place, make sure that any crown in the board is facing up; in time, gravity and the weight of the decking will straighten the joists.

There are at least two good techniques for attaching joists to the ledger, and at least one that should not be used. One reason many decks decay at this location is that the joists are toenailed into the ledger; nails split the ends of the joists, allowing water to collect exactly where it shouldn't. Joist hangers minimize splitting at the joist end, and will prevent water from getting trapped in this crucial joint. We first nail the hangers to the joists, and then fit the assembly to a level chalkline. This compensates for slight variations in joist width.

A cleat can also be used to support the ends of the joists. You still need to toenail the joists, but splitting is reduced because the nails can be a smaller size since they do not carry the weight of the deck. If you use this technique, the ledger should be one dimension wider than the joists. For example, use a 2x8 ledger with 2x6 joists. This will allow room for a 2x2 cleat to be attached to the ledger. Run a bead of silicone caulk along the cleat/ledger seam to keep the water out.

Decking—If the decking surface is not applied properly, it will be the first thing to deteriorate, causing a chain reaction of decay throughout

Anatomy of a deck

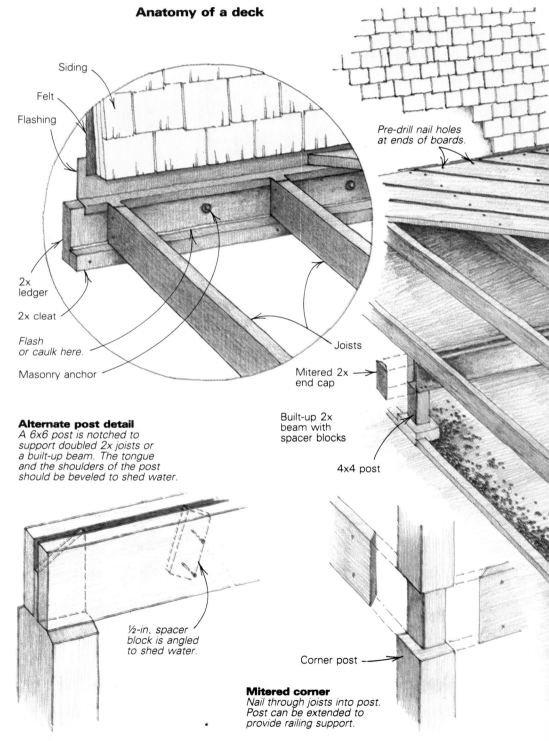

Siding

Felt

Flashing

2x ledger

2x cleat

Flash or caulk here.

Masonry anchor

Pre-drill nail holes at ends of boards.

Joists

Mitered 2x end cap

Built-up 2x beam with spacer blocks

4x4 post

Alternate post detail
A 6x6 post is notched to support doubled 2x joists or a built-up beam. The tongue and the shoulders of the post should be beveled to shed water.

½-in. spacer block is angled to shed water.

Corner post

Mitered corner
Nail through joists into post. Post can be extended to provide railing support.

Drawing: Frances Ashforth

the rest of the structure. *Never* butt the ends of two decking members together and nail into a single joist. This is one of the major causes of decking failure (top photo, next page). The reason is that water collects in the seam between the butting boards and enters their end grain. And when two boards butt over a single joist, the problem is intensified. With only ¾ in. for each deck board to be nailed into, the nails must be placed very close to the end of the board, encouraging splitting. We get around these problems by doubling up strategic joists, using a block of wood between them to create a 1½-in. space. The end of each deck board cantilevers over this space so water can't collect. This also allows the boards to be nailed farther from their ends, which minimizes splitting. If

Bench supports built from 2x8s can be trimmed to a width of about 3½ in. where they form the backrest. The supports should be securely nailed or bolted to the deck's supporting structure.

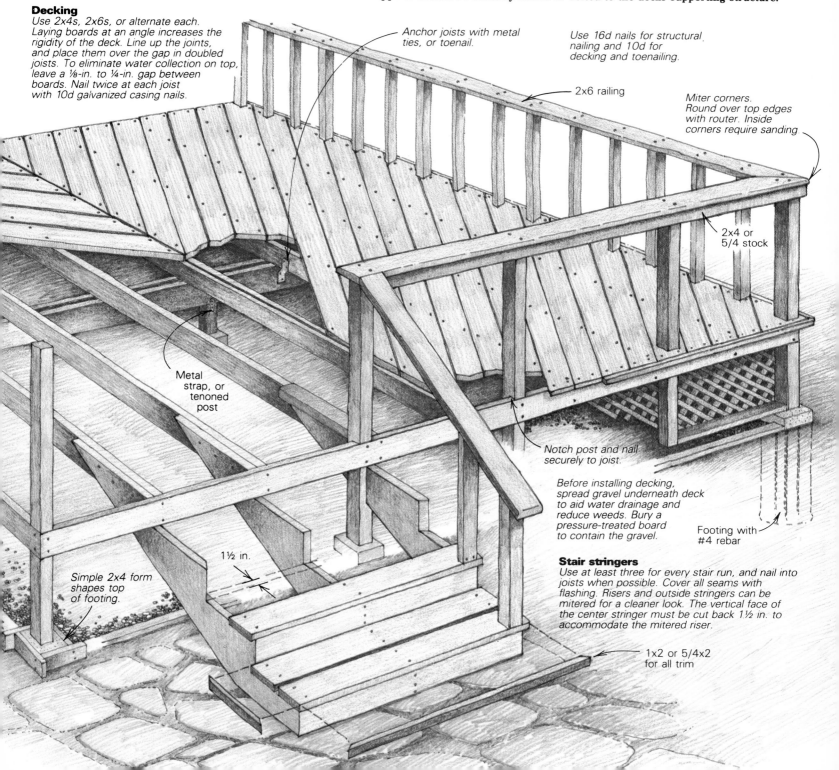

Decking
Use 2x4s, 2x6s, or alternate each. Laying boards at an angle increases the rigidity of the deck. Line up the joints, and place them over the gap in doubled joists. To eliminate water collection on top, leave a ⅛-in. to ¼-in. gap between boards. Nail twice at each joist with 10d galvanized casing nails.

Anchor joists with metal ties, or toenail.

Use 16d nails for structural nailing and 10d for decking and toenailing.

2x6 railing

Miter corners. Round over top edges with router. Inside corners require sanding.

2x4 or 5/4 stock

Metal strap, or tenoned post

Notch post and nail securely to joist.

Before installing decking, spread gravel underneath deck to aid water drainage and reduce weeds. Bury a pressure-treated board to contain the gravel.

Footing with #4 rebar

Simple 2x4 form shapes top of footing.

1½ in.

Stair stringers
Use at least three for every stair run, and nail into joists when possible. Cover all seams with flashing. Risers and outside stringers can be mitered for a cleaner look. The vertical face of the center stringer must be cut back 1½ in. to accommodate the mitered riser.

1x2 or 5/4x2 for all trim

The premautre failure of decking is often caused by nailing too close to the end of the board. This encourages splitting, and allows water to accumulate around the board ends.

Cutting the decking to length after nailing it in place ensures a clean, uniform edge and saves time otherwise spent cutting boards one by one. Run the boards long and snap a chalkline to mark the cutting path, or use a straightedge to guide the shoe of the saw. A built-up beam and a mitered joist corner can be seen in this photo. Wherever joists along the perimeter of the deck meet, a mitered connection protects end grain from direct exposure to the weather.

you must nail closer than 2 in. to the end of a board, predrilling the nail holes can also help to reduce splitting.

For the decking surface, we like to use 2x6s or sometimes 5/4 by 6s if the quality is good. You can also use 2x4s, but you'll have a lot more nailing to do. If you have the choice, nail the decking cup-side down to prevent any water from pooling on the individual boards.

We like to space the decking boards about ⅛ in. apart (the thickness of a 10d nail). The boards will shrink, depending on their moisture content, and we have seen this space expand up to ¼ in. If there are a lot of deciduous trees with small leaves close to the deck, you might want a wider spacing to allow the leaves to fall through. When you're installing the decking, some boards will most likely be crooked. With a flat bar and some lever action, one person can easily straighten out each board while nailing.

Stairways—Stairways can be dangerous areas, and require special attention. Codes usually call for at least one railing at the side of the stairs if they include more than two risers. Wide stairways have a spacious and inviting appearance, so we like to build them at least 4 ft. wide, enough for two people to pass comfortably.

Inconsistency in the height of a step or length of a tread is dangerous and awkward. We have found that a shorter rise and longer tread is easier to walk up, safer and very elegant. A rise of about 7 in. seems to be make a comfortable step, and allows us to use an untrimmed 2x6 for the risers. Sometimes we miter the riser boards into the outside stringers. We're not fond of exposed stringers. We use a pair of 2x6s for the tread; these make a run of a bit more than 12 in. with a trim board.

We use three stringers for deck stairs, even if the stairway is only 30 in. wide. We have seen too many stairways warp and fall apart with only two stringers. Using three-stringer construction isn't too difficult and doesn't cost much more. In fact, the trickiest part of building one is cutting each stringer to identical shape and lining them all up in the same plane on the rise and run. You can buy framing clips (called stair-gauge buttons) to keep your framing square in the right position while you lay out each stringer, but this doesn't always get you past Murphy's law. Our trick is to lay out one stringer only and clamp it to another length of stringer stock. When you cut the first one, the sawblade scores the second one, saving you one layout. Repeat the process with additional stringers. This will duplicate all the stringers and save layout time.

Railings—Railings are important to the safety and appearance of decks. As a design element, they can be the highlight and outline of your project. Railings are one of the places where a designer's creativity can be expressed, and there is no one way to build them. But there are some general rules to follow.

The most important characteristic of a good railing is strength. You can be sure that people will lean against the railing, and often they will sit on it, too. A strong railing is particularly important on elevated decks. We like to play it

safe, and build our railings as strong as possible. To do this, we solidly attach the posts or balusters to the deck joists. These structural supports should be located no more than 48 in. apart, and the railing should be about 34 in. above the decking surface. Be sure that the edges of the railing are well sanded, especially in places that will get a lot of use, like stair railings.

Consider the spacing between your railing uprights as part of the project's visual design. Close spacing visually encloses the space, and also prevents small children from falling through. Railings with fewer uprights will visually expand the space and be less inhibiting to your view.

One type of railing we use combines a bevel-cut 2x6 and a matching 2x4 into an "L" shape. We position the 2x6 horizontally, and our clients enjoy the strong visual effect this creates. The 2x6 acts as a cap for the railing uprights, shielding their end grain from the elements. Although tricky, this technique allows for some very interesting joinery at all corners. To clean up the mitered edge, we round it over with a router or belt sander. A simpler railing is shown in the drawing, previous page.

Seating—Built-in seating is a great way to finish the deck. As with railings, there are many ways to build it, and no one way is correct.

You can build seating either with or without a backrest, and the choice will often depend on whether or not an unobstructed view is important. If comfort is more important, you'll want to build at least some of the seats with a backrest. The top of the backrest should be between 30 in. and 34 in. from the deck, which can be designed nicely to tie in with the railing.

A backless bench can function as a physical barrier for a deck edge without acting as a visual barrier as well. We have also used a low, wide railing as a mini-bench, which also makes a good place to display potted plants. And sometimes we'll use a built-in planter to serve as a visual barrier.

Benches can be difficult to build, since they not only have to be strong, but comfortable too. With backless benches, we have used 4x4s or 2x6s as supports. For a bench with a backrest, we use a 2x8 as the seat support, and rip it down to a 2x4 for the backrest support (photo previous page). A 15° backward lean seems to be comfortable. We then run our mitered railing across the top at 32 in. The cross supports for the seat are 2x6s cut into long, wide triangles. For the seat, we use three 2x6s with a 2x4 band. This will give the seat a total width of 18 in. The standard seat height that we use is 17 in. In calculating seating height, don't forget to account for any cushions that you might use. Save your best boards for construction of the seating, because this part of the deck will be well used and very visible.

A final note—Use these tips in combination with your local codes. With a little creativity and your basic construction knowledge, the deck you build should last a long time. □

Scott Grove is a partner in Effective Design, a design/build company in Rochester, N. Y.

Photos: Scott Grove

Carved ridgepole

The 12x20 redwood ridgepole on a house designed and built by David Clayton extends about 4 ft. beyond the rafters. The huge timber, salvaged from an old railroad bridge, was carved in place by Jose Cross. The life-size king salmon overlooks the Mendocino (Calif.) coastline.

Cedar post

Shortened by several feet and only slightly relocated from the nearby woods, the trunk of a cedar tree supports the 6x10 ridgepole of a post-and-beam addition to an older house. The owners cut the tree trunk to size and set it on the subfloor, directly above a cement-block wall in the basement. The oak plank floor extends to an octagonal border. The trunk is surrounded by stones and moss, an Oriental touch that makes it seem as if the tree had grown there. The untrimmed upper branches continue the theme.

—*Thierry Guerlain, Redding, Conn.*

Rebar pin
through ridgepole
to tree trunk

Trunk
toenailed
to joists

Stones
and moss

Border

Subfloor

Cement-block
wall
to footing

If you enjoyed this book, you're going to love our magazine.

A year's subscription to *Fine Homebuilding* brings you the kind of practical, hands-on information you found in this book and much more. In issue after issue—seven times a year—you'll find projects that teach new skills, demonstrations of tools and techniques, design ideas, and new materials and methods that will bring home building into the next decade. You'll also find articles about historically significant architects and designers and the contributions they've made to home building.

To subscribe, just fill out one of the attached subscription cards, or call us toll-free at **1-800-888-8286.** And as always, **we guarantee your satisfaction.**

Subscribe Today!

7 issues for just $26

The Taunton Press
63 S. Main Street, Box 5506, Newtown, CT 06470-5506

Fine Homebuilding

Use this card to subscribe or to request additional information about Taunton Press magazines, books and videos.

☐ **1 year (7 issues) for just $26**—over $8 off the newsstand price
Outside the U.S. $30/year (U.S. funds, please)

☐ **2 years (14 issues) for just $47**—over $22 off the newsstand price
Outside the U.S. $53/year (U.S. funds, please)

☐ **3 years (21 issues) for just $66**—over $37 off the newsstand price
Outside the U.S. $72/year (U.S. funds, please)

Name _____

Address _____

City _____

State _____ Zip _____

☐ My payment is enclosed. ☐ Please bill me.

☐ Please send me information about other Taunton Press magazines and books.

☐ Please send me information about *Fine Homebuilding* videotapes.

Fine Homebuilding

Use this card to subscribe or to request additional information about Taunton Press magazines, books and videos.

☐ **1 year (7 issues) for just $26**—over $8 off the newsstand price
Outside the U.S. $30/year (U.S. funds, please)

☐ **2 years (14 issues) for just $47**—over $22 off the newsstand price
Outside the U.S. $53/year (U.S. funds, please)

☐ **3 years (21 issues) for just $66**—over $37 off the newsstand price
Outside the U.S. $72/year (U.S. funds, please)

Name _____

Address _____

City _____

State _____ Zip _____

☐ My payment is enclosed. ☐ Please bill me.

☐ Please send me information about other Taunton Press magazines and books.

☐ Please send me information about *Fine Homebuilding* videotapes.

BUSINESS REPLY MAIL
FIRST CLASS PERMIT No. 19 NEWTOWN, CT

POSTAGE WILL BE PAID BY ADDRESSEE

TheTauntonPress

63 South Main Street

Box 9974

Newtown, CT 06470-9974

NO POSTAGE
NECESSARY
IF MAILED
IN THE
UNITED STATES